ROCKITECTURE

The Symphony of River Rocks and the Men Who Listened to Their Music
Indigenous Architecture in California 1885–1935

By Barry Schweiger, AIAE
Principal Photographer–Scott Clark, Additional Photographs–Guy Eisner,
and Historical Photos

Barry J. Schweiger
www.barrywritings.com

ISBN Hardcover 978-1-7379733-0-0
ISBN Paperback 978-1-7379733-1-7
ISBN E-Book 978-1-7379733-2-4

Book design by Rick Schank of Purple Couch Creative
Cover photo retouching by Kevin O'Neill

Printed in the United States of America

DEDICATION

To all those who boarded a new train to travel the expanses of the wide plains
and endure the wider still great deserts of the Southwest
to arrive in a booming town of oranges and real estate and oil and movies.
My mother was one of those fearless people.

ADDITIONAL DEDICATION

It closed the world, normal activities stopped.
Hospitals overflowed with COVID-19 patients.
And morgues were full.
2019-2020, while writing the final chapters, became the years that suppressed our lives.
Research and interviews stopped in the lock downs.
And then, more sudden than the beginning of the pandemic, another tragedy.
My companion of twenty-six years died, instantly.
Our empty house gave no console.
In those grieving months, I am grateful for Jill, my editor,
Scott, our photographer, my family and life-long friends for encouraging me to complete the final chapters.
"Donna was proud of you and wanted you to
finish this book," they told me.
And to her I dedicate all our work.

CONTENTS

INTRODUCTION

"The loss of Venice," Jeff Goodall writes, "wouldn't be about just the loss of present-day Venetians. It's the loss of the stones in the narrow streets where Titian and Giorgione walked. It's the loss of eleventh century mosaics. The loss of Venice is about the loss of a part of ourselves that reaches back in time and binds us together as civilized people."

Bill McKibben, *Falter*, p. 45, Henry Holt and Co.

We searched for eight years and looked for the who, what, where, when, why, and how of these stunning Southern California cobblestone homes. We thought that many would be interested and enjoy this particular piece of unrecorded California history, California's Stone Age of the early twentieth century, and performed this effort to share these wonders. We learned much of California's history, its geology, and the dreams, greed, and of course hubris of the settlers in their desperate search for health and wealth. Though very little history of theirs was recorded, we offer a few insights and some fascinating stories. We admire and love these stone buildings and do not want to lose this indigenous vernacular or ignore it.

For decades we've traveled most of California, though we haven't covered but a fraction of this large state, roaming from its sandy deserted deserts to the sandy crowded beaches, climbed its 11,000-foot snow-covered mountain peaks, and visited the Northern California coastal redwood forests. Hiking, camping, driving, and moteling through California, we found river rock buildings scattered, one of the state's many hidden treasures.

A stone castle in Forest Falls, San Gabriel Mountains

But the river rock homes stood only in particular locations, seemingly hidden to history, hidden to our daily lives, yet loved by their owners. Any attempt to appreciate the river rock structures must begin with yesterday's mirrors as these structures symbolize California's birth and its maturing years. No one person or group of people can lay claim to this style, for it is indigenous, of the location. These buildings resulted from the efforts and dreams of multitudes only during a few decades, long ago, never to be repeated. But the buildings can be respected and preserved for those yet to come.

River rock homes fascinate. With the magic of childhood imagination, for me they would change into powerful medieval castles. As a child living in the small town of Sierra Madre located at the base of the mighty San Gabriel Mountains, I dreamt of being a fearless knight living in a stone home, my castle. Imagination transformed these stone homes into crenellated forts, with front yard lawns changing into green water-filled moats and the driveways turning into lowered-drawbridges where knights in clanging metal armor charged to the battlefields on their powerful horses whose large hoofs pounded the vibrating wooden bridge. Strong and impenetrable, their interiors dark as closets, stone buildings filled me with fantasy.

2123 Crescent Avenue, Montrose

Playing in the granite foothills, filling my pockets with stones, I learned to love rocks, their colors, their sparkling, their rough texture and rounded shapes. And best yet, it thrilled me that people used these stones to build homes and churches.

But instead of studying geology, the stones led me to architecture, where I learned harmony, balance, and proportion. Even after years of study and decades of professional work in Southern California, my passion for those indigenous stone homes never waned. No matter where I went, they lured me.

This book discovers and uncovers the long-forgotten river rock homes of Southern California, along with some of their creators and their stories. They present a universe of history, geology, sociology, economics, masonry technique, and stories taking us back millions of years to the formative times of Southern California's monolithic granitic mountains, to the land's earthquakes, to how river rocks came to be lying in the dry canyon washes and in the dried compacted mud of alluvial fans. Unused by the native Tujunga and Gabrielino tribes for their homes, the Spanish for their missions, or the Mexicans in their rancheros, the river rocks waited for the settlers from the United States, and for an invention. Theirs is a long lost era. The settlers knew the earth, knew beauty, knew the truth in nature and knew simplicity. With an appreciation of history, I have sought to bring the once-mysterious past into the present. During our search for river rock homes I became aware that our lives would not exist without the lives of our ancestors. Our stories do not exist independent of their stories. Context became critical to understanding the stone homes of Southern California.

Lacking forests, SoCal in the early years of its settlement, imported redwood from the forests in Northern California and Oregon, which was milled and then shipped down the coast to piers and harbors in Southern California. As opposed to the river stocks, stones sat there in the warm washes of the San Gabriel Mountains, readily available as native building material to be used by only a few. Constructing with stone is not a recent technology; it had been used for thousands of years, but never in California until the early twentieth century.

In his new SUV, my friend Scott and I traveled the by-roads for years. We drove from the sands of the Pacific Ocean to the sands of the Mojave Desert, from the heat at the Mexican border to the cool air of Sonoma County where we discovered way-side communities, pockets of repose hidden in the city's corners. We learned geology and heard interesting stories all in search of California's river rock buildings, enough that two books are needed, with this being the first.

Stories about these river rock homes are limited, although California's history is well documented. River rock buildings in California were a one-time event, though in more recent years their historic preservation has returned in many communities. The stone homes, nestled in the San Gabriel Mountain foothills, which this book highlights, have stood since the days of our great-grandparents and, when shielded from wanton demolition, will outlive our great grandchildren. As Henry Wordsworth Longfellow said so succinctly, they remained "untouched by history."

And like these buildings of blended cobblestones, history also presents a blend of parts that when viewed again becomes a whole of interacting parts, a synergy. Our lives would not exist without the lives of our ancestors.

Arriving in SoCal over a century ago, the settlers of this vacant foothill land sought health, wealth, and opportunity and chose the sloping foothills of the San Gabriel Mountains over the flat lands of Los Angeles. In those early years in the rural areas of Southern California, settlers found granite river rocks and used those stones to build their homes. These rare buildings came into being in California during the early twentieth century, when the latest technologies included the telephone, radio, electricity, indoor plumbing, special kitchen gadgets, and the Model T-Ford. Yet even with that burgeoning technology, the new settlers built with stone, an ancient craft. Over time, these buildings rested no longer on open land, but became crowded by bland and uninspiring city landscapes and its endless asphalt; yet, they still awaken our eyes and our souls with their symphonic music of visual texture, wonderment, and the mason-craftsman's aesthetic skills. Contrasted to current trends that use inexperienced people to produce uninspiring buildings, the master masons years ago gave us inspiration with their stone art.

Drive by (or read) and stop, admire, study how the masons built these buildings. How did the masons learn the art of a river rock building that withstands earthquakes? Their work resembles trying to stack marbles, or basketballs, or making an arch with melons. How were these masons of the early twentieth century considered? How did society consider their buildings? And how does today's culture value them?

Seen as unique and exceptional today, these stone homes when built were considered routine, or at least not exceptional. Newspaper articles and letters would refer to someone building a home of stone, without further elaboration. No why, no discussion, no big thing. Yet a century later they have become a legacy to the settlement of Southern California in the early twentieth century.

A few of those settlers were experienced masons from England, Ireland, Italy, France, and Germany, while others learned their trade from these European immigrants, and others were self-taught. Then in the mid-1930s California building codes changed, effectively halting the stone masons from creating these distinctive structures, just as their artistic and masonry skills had developed and matured. Since then, river rock buildings have become rare and historically valuable.

It took years' long effort and research to find them, and while writing this book, the COVID-19 pandemic enveloped the world stopping all but essential activities, thus stopping our in-depth field research. I found it eerily coincidental that we would be writing of the Spanish Flu outbreak of 1918, which were the years some of these buildings were constructed, while sheltering in place and wearing face masks during the 2020-21 pandemic. A century had elapsed since that first pandemic, and we expected that medical progress would shield us from such viruses; today as then the medical profession searches for cures and politicians delay and debate, while people die. This time it may have begun in Wuhan, China; while in 1918 the pandemic may have started at a pig farm in Kansas. Back then it contributed to the forces that pushed for much-needed universal health care in European countries. But not in the United States, even though 675,000 Americans died from the deadly flu back then. We hope, with faith in science, that this time will be different.

"It's deja vu all over again," Yogi Berra, the New York Yankees catcher, said. A deadly Black Swan virus returned a century later, and libraries closed, museums closed, researched stopped, but I continued to write, leaving some chapters sketchy and missing details.

Some of the remaining stone homes are derelict, abandoned and pleading for renewal; some had been restored and cared for; and some had been demolished, taken by the economic forces of rezoning for increased density, or the building of freeways by government agencies apparently unappreciative of this rare indigenous architecture. All hold unique stories and all are located on lands of particular geology.

Undoubtedly there are many more hidden and unacknowledged treasures that we did not find. Gems unknown to all except the 'locals'. They represent California's contribution to the refinement of river rock art and its technology. Some of these stone buildings became Historic Sites due to their being constructed of river rocks by famous people. But most were overlooked. Why weren't more built? Certainly the stones remained available. Stone homes in California are as rare as the California condor. We'll surmise a few thoughts on this.

Construction of these stone buildings occurred over an abbreviated period of time, between 1885 and 1935, with one famous outlier constructed during the 1950s. They present an architectural collection of styles and sizes. Why build of cobblestones? Where do they come from? How did they became round smooth shapes, and who were these early Southern Californians who built them? Looking into the seams of our unimaginative and ersatz urban milieu we found a few answers in the artful craftsmanship of this indigenous architecture.

Join us on this journey of *Rockitecture*, experience the symphony of the river rocks, and learn of the men who listened to their music. Bring a chair, a snack, your camera and absorb their enduring beauty. What would you have used for your new home at the foothills of Southern California mountains? Cobblestone or wood? Viewing these enhances our awareness of their textured complexity, like a star-lit sky, simple yet mesmerizing beauty, and to enhance our appreciation of our ancestors, ancestors who not only built them but those who lived in a cobblestone home. Living in a thick-walled river rock home may rekindle that long time ago when our ancestors lived in thick protective shield of caves and castles.

Yet in disclosing these works of granite art, I felt mixed emotions. My motivating intentions were to share their beauty and possibly awaken SoCal's historic societies and planning commissions to these jewels. Disturbing their occupants or subjecting vacant buildings to vandalism was never intended, and we do hope this work brings pride to the owners and their communities for they occupy a unique and special piece of California history.

And now we visit the hallowed stones of St. Luke's of the Mountains Church in Tujunga.

A poem from an old farmer's notebook

THE STONES

I owned a slope full of stones
Like buried pianos they lay buried in the ground
Shards of old sea ledges, stumbling blocks
Where the earth caught and kept them
dark, an old music mute in them
that my head keeps now that I have dug them out.
I broke them where they slugged in their dark
cells, and lifted them up in pieces.
As I piled them in the light
I began their music. I heard their old lime
rouse in breath of song that has not left me.
I gave pain and weariness to their bearing out.
What bond I have made with the earth,
having worn myself against it? It is a fatal singing
I have carried with out of that day.
The stones have given me music
that figures for me their holes in the earth
and their long lying in them dark.
They have have taught me the weariness that loves the ground,
and I must prepare a fitting silence.

CHAPTER 1

The Search Begins

"The population of these foothill towns could fit in the Rose Bowl then."

"It is good to visit a spot like this, for the 'sermons in stones,' that one may find here are many." [1]

Golden Gate Bridge, Notre Dame Cathedral, St. Louis's Gateway Arch, Statue of Liberty, Tujunga's St. Luke's of the Mountains—architecture defines a city, speaks of its character and its history. During our search for the river rock homes of Southern California, we met people who knew little of their home, unaware of its uniqueness, and less about the people who built it. In our speedy life we skip over our predecessors' creations and their history.

Should you have passed over reading the introduction, I encourage to go back and do so for there we learn a little of the context and scope of these rare buildings and their communities.

Throughout the state are by-ways, small towns, seldom-used roads, and places of history lying quiet waiting to be discovered. "It is good," C.F. Carter tells us in *Some By-Ways of California* "for the 'sermons in stone' that one may find many."[2]

And those stone sermons enrich life. I admit that I'm biased; beauty enlivens and history enlightens. Being an architect for over forty years taught me the intrinsic life-giving value of proportion, textures, craftsmanship, and in situ, the in-place environment and culture of a building's neighborhood. And that is what these cobblestone buildings offer: local human and geologic histories which may provide beneficial guidance.

An example, building in the foothills of the San Gabriel Mountains of Southern California meant

How did the builders and owners see their new cobblestone home? As a return to nature from the clamor and soot of industrialized cities? As sound shelter or as status symbols? Were these homes visual expression of wealth or looked upon as low income houses?

A well-preserved example of California's cobblestone buildings, Our Lady of Guadalupe Mission in Irwindale.

becoming neighbors with the original inhabitants: mountain lions, rattlesnakes, rabbits, and coyotes, along with earthquakes, landslides, fires, and floods. That may be okay with some, but not many. Especially those with small pets or porcelain ware. But most learn the unfortunate way, as Portola and the Spanish settlers learned of Southern California's character the hard way.

In 1776, a memorable date far away on the east coast of America in Philadelphia, and five years after Portola's small party of two Franciscan priests and ten soldiers established the first adobe mission by the Santa Ana River, naming it the San Gabriel Mission, a great flood destroyed it. Their second attempt at living with Southern California's tempestuous climate succeeded. They rebuilt five miles to the north, still of adobe for that was the technology then, where it remains an active church and community relic. (Yet while writing this an accidental fire started during repair work and consumed a large portion of the wood roof and its two-hundred-fifty-year-old interior.)

Many of Southern California's cobblestone buildings hold similar stories of fire, flood, or demolition and many remained unscathed, loved, and preserved by the local residents. An example of the threatened to be destroyed category resides in Tujunga, there a well-preserved stone church constructed a hundred years ago became threatened by a religious bureaucracy.

A NEIGHBORHOOD STORY

After the morning services at St. Luke's church were completed and the parishioners returned to their homes either in their car or strolling enjoying the clear sunny weather, we met Stuart Byles in the stone church's patio. The unofficial community historian, Byles related a little history, "Yes, they were goin' to tear her down." He looked at the ground as if reliving those threatening days. Byles is a retired general contractor, full of community pride, and a soft-spoken yet enthusiastic historian of his Cresenta Valley, which is the home of St. Luke's church and many stone homes. Cresenta Valley is long and narrow sandwiched between the youthful and still growing Verdugo Mountains on its south border and the enormous adult, also still growing with each earthquake, San Gabriel Mountains on the north border. Over millennia, the valley became filled with storm-eroded rock and mud from multiple canyons blending to create an enormous alluvial fan (see chapter 2) of compacted erosion. This six-mile-

St. Luke's of the Mountains cobblestone Church, Tujunga, Cresenta Valley.

long valley accumulated one of the densest collection of cobblestones in Southern California and became the site of dozens of stone homes.

Sitting on a stone wall in the church patio, hands clasped between his legs, Byles, remembering those contentious days, swayed his head from side to side. Sighing, he looked up. Small patches of late morning sunlight filtered through centuries-old oak trees onto his unlined and sun-browned face. "We fought back," he said. "Told the diocese they would not get a penny from our congregation for any new church. This one's our center, and she stays."

Built in the 1920s the church was based on a local artist's painting of a stone chapel he saw on a visit to France's countryside; St. Luke's continues unaffected and undisturbed by many tumultuous decades of city growth, war, depressions, and earthquakes. Today it appears to be young, just a few years old. Time doesn't affect stone construction. Situated on a moderate slope among ancient California oak and pepper trees, the stone church's calming influence contrasts with its chaotic neighbor, a noisy fast moving four-lane highway, appropriately named Foothill Boulevard.

Seymore Thomas' painting of a rural chapel in France, the inspiration used to build St. Luke's.

Here in the valley of La Cresenta, stones lie buried in the alluvial soil and in the nearby dry Arroyo Seco canyon to the east. La Cresenta Valley drained into and helped carve out the Arroyo making it wide enough to later fit the Rose Bowl and a golf course. The Arroyo and the Valley supplied the church's granite stones. I wondered how these rounded stones occurred here, why the settlers used the cobblestone for their buildings, and who the masons were that created these unique buildings?

"Stuart," I said, "we've researched libraries and historical societies and found not one publication, not one book that covered these river rock buildings. Also, we searched for a book on this unique church and its construction. I didn't find anything."

"I can take you to the rock buildings around here, but, you know, I'm not aware of any such book either.

Sorry...can't help you there."

That this historical vernacular architecture hasn't been documented revealed a significant oversight. These modest and indigenous stone buildings, built in the early twentieth century, seemed to be lacking appreciation and in need of protection – from ourselves, from encroaching highways, freeways, and unrelenting development. A prescient and observant 1979 building survey of California warned that "They (the stone buildings) call for attention that historical preservation programs provide." [3] Over the years this warning was ignored.

In the beginning, it appeared that compiling a book on the stone homes of California would be a moderate undertaking. Yet, eight years later this endeavor had taken a life of its own, becoming enlarged, complicated, and demanding ever more research. And that endless search began with a few cobblestone homes here in the San Gabriel Mountain foothills, but then expanded to discover over two-hundred California stone homes. Two volumes were needed.

Book I focuses on the stone homes in Southern California and Book II will include California from the Mojave Desert to the wine country of Sonoma County.

A historical postcard photo when ivy grew on the stone bell tower and Easter lilies bloomed in the front yard.

In our effort to record these ignored masterpieces, photographer friend Scott Clark and I drove California's backroads and urban highways, visited wayside towns and neighborhoods, communities that remain forgotten, and places once forgotten yet have become reinvigorated with the vibrancy of new generations. At each location, volunteers at the community's historical societies enthusiastically shared their stories and materials with us. We researched the history and geology of Southern California to find context. We photographed the stone homes to capture their texture and beauty for historians, lovers of architecture, their proud owners, and future generations.

How does it feel to be surrounded by eighteen to thirty inch thick granite walls? Are the interiors cool or warm? When inside does it feel like living in a mausoleum? Do you feel more secure in a stone home than a stick built home? The fact that history recorded nothing on the these buildings it appeared that, to the settlers, rock homes were not unusual. Well, they are now—for their unrepeatable uniqueness and the visual sensuousness of their indigenous cobblestone architecture.

Continuing, Byles said, "Our stone church was our community anchor, we told 'em. Not only for the church services, weddings, and socials...where we serve cakes and mac-n-cheese, but just because it is here... solid like... reminding us of God's creation with the beauty of these river stones."

"No new church," he concluded, "of wood and plaster can match that." And upon hearing this, I suspected that the settlers and masons a hundred years ago would understand this sentiment. Outlasting earthquakes and threats, St. Luke's of the Mountains remains. Others haven't.

East of downtown Los Angeles in the area of Fetterly Ave. shows the terrain and landscape to which the immigrants relocated. Much work was needed to be done by energetic and visionary people.

THEIR CULTURAL ENVIRONMENT

Understanding and appreciating these cobblestone buildings means showing their original setting–

undeveloped barren land. And their cultural and technological context. For it is difficult to see the value these buildings brought the settlers through today's eyes. Visit museums that focus on early 1900s Southern California and learn of the settlers struggles and their joys remembering that what we are accustomed to did not exist then. Yet their lives where nonetheless rich for it.

Much is written about the City of Los Angeles, which became the economic and cultural hub of SoCal, so this will be a very brief and quick overview. Later we'll learn about the people who immigrated to SoCal along with a few ideas on the settlers' characters, those who traveled thousands of mile to this desert 'land of little rain.'[4]

Within forty years this barren land of dry native grasses and stony soil changed into a developed land of 2.2 million people. And the people came because they wanted to live in this land full of promises and dreams, where visions of untrammeled prosperity and miraculous health could be realized.

"The Great War ended and the Great Migration began," Bruce Henstell wrote on Los Angeles history. "Out of the Midwestern farms, out of the dreary industrial cities of the East, across the plains and over the desert

Promotional pamphlet used to lure people to settle in SoCal and buy a plot of land from the developer. Lack of water would soon become a problem.

The Los Angeles Times *bicycle club on an outing on unpaved Western Avenue near Pico Boulevard. This historical photo from the 1890s shows the Santa Monica Hills in the background.*

they came, a trek if not overly perilous at least laborious.... What dream drove them? The promise of sunshine and wealth...easy three-day access via the Southern Pacific Railroad...At the end of the rainbow was Oz in an orange grove along with a white cottage and a detached garage with a late model Ford. What more was there to life?" [5]

Yet many others arrived to open this undeveloped land with its scattering of sun-ripened vineyards, a few ranchos, and citrus farms.

In the late 1800s and early 1900s, those immigrants built homes with wood imported from the redwood forests of northern California and Oregon while their office buildings relied on brick – construction methods imported from the eastern U. S. They only knew these types of construction; they were familiar with them, not the adobe and wood construction of the land's prior occupants. In these years Los Angeles was to be transformed into a teeming city and Southern California sprouted dozens of towns and neighborhoods, some based on agriculture and some based on commerce and finance. Later in the 1920s, around one hundred thousand people came each year. Clerks and merchants left their jobs and instantly became real-estate agents selling newly developed lots. In his excellently researched opus on Los Angeles, John Caughey notes that "Building permits rose from $28 million in 1919 to $121 million in 1922. In those three years building in Los Angeles exploded by a million dollars a year or 432%. A year later in 1923 the grand total was $200 million."[6] That becomes an incredible 714% over four years. And this was only in the city of Los Angeles, not the whole of Southern California where many areas and towns did not keep records.

Children play on Western Avenue south of Sunset, circa 1906. Photo courtesy of the USC Libraries – California Historical Society Collection.

And during that period a few cities kept track of building though, by issuing permits. We found that of the approximately two hundred river rock buildings we identified one hundred and forty eight were issued permits: from 1900 to 1915, twenty-one river rock buildings were constructed; from 1916 to 1929 ninety-five; and during 1930 to 1945 thirty-two. This might indicate that to build a stone home was not only practical but also required above average capital.

Newspapers reported on the day's significant events, such as in 1928 when the Franciscito Dam on the Santa Clara River failed killing four hundred residents in a massive flood that continued racing miles westward down the Santa Clarita valley and across Oxnard's vast farm lands to the Pacific Ocean. A year later the papers reported on the stock market crash, and four years later on the earthquake centered on the Newport-Inglewood Fault destroying Long Beach and much of the surrounding cities.

Downtown L.A. 1920s, population of Los Angeles in 1880 = 11,311; in 1900 - 102,000; in 1910 = 504,131

Southern California's turbulent nature continued the following year, for in the winter of 1933 forest fires burned uncontrolled across the San Gabriel Mountains and then torrential rains produced massive floods devastating the foothill communities, and this was to be repeated in 1938 with "Torrential days long floods,...two of the worst on record."[7] But an observer of those years writes, "in the thirties Los Angeles was spacious an uncluttered... unmarred by crowds or polluted air."[8] Yet many early immigrants must have wondered why they moved to this turbulent land.

The old map on the following page shows all these new towns in early twentieth century who endured these natural hardships. During the decade of the 1920s, our cobblestone buildings began to appear in the foothill towns of San Fernando extending east to the towns of Cucamonga and Upland by the Cajon Pass, small fledging towns that survived fires, floods and earthquakes.

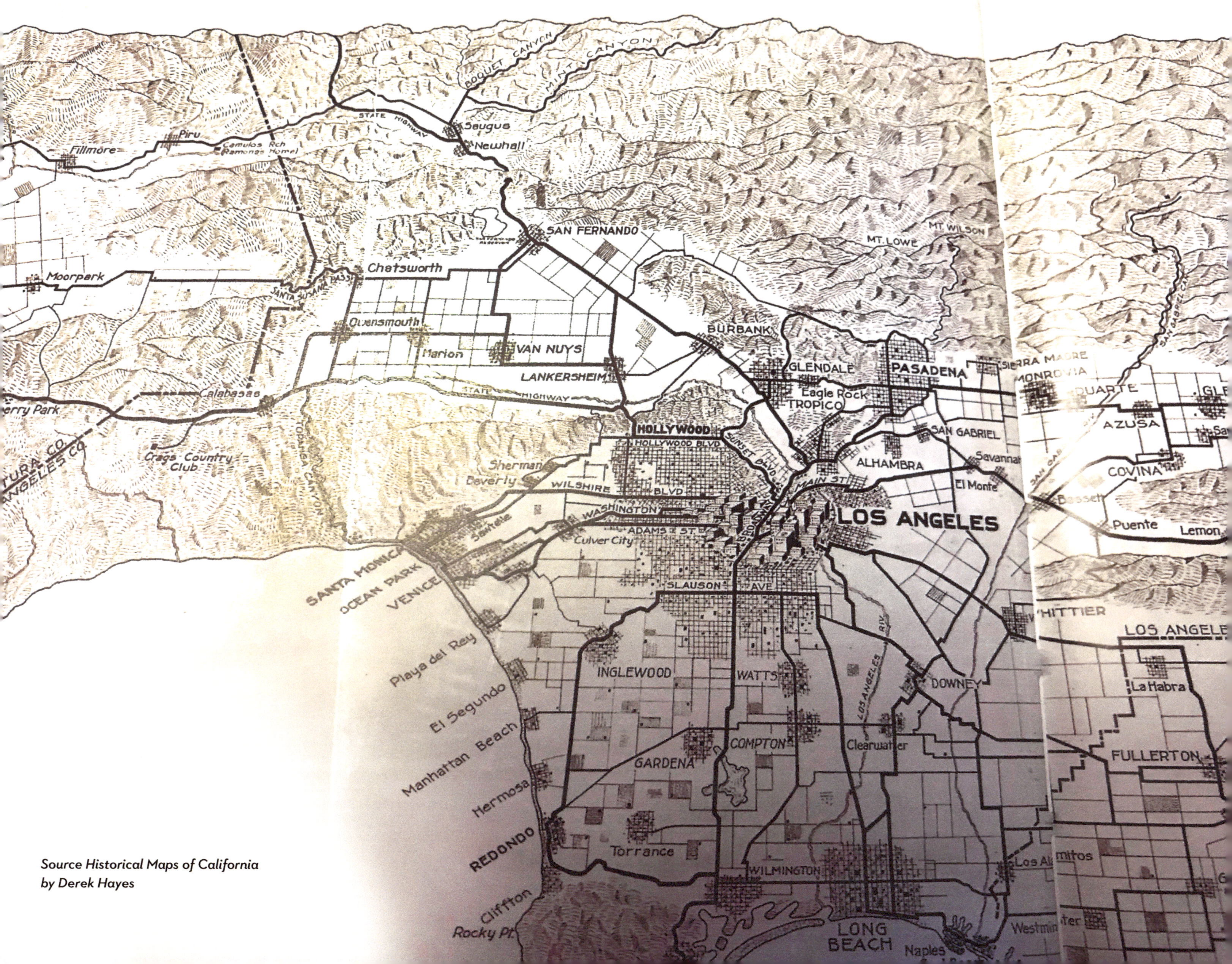

Source Historical Maps of California by Derek Hayes

Los Angeles Oil Boom, photo from book **Los Angeles, An Illustrated History, *p. 137***

"The story of California has been the story of land and water." [9] And oil. Geologist John Dvorak relates that oil was once considered a nuisance as wagons and cows got stuck in the gooey stuff. Though "By 1925, California produced almost a quarter of the world's oil - and production soon outstripped the demand."[10]

Southern California's settlers arrived on lands "the Indians of California bequeathed...a paradise they had been stewards of for 10,000 years."[11] They made new towns in the valleys that the Indians had named 'The land of smokes', which described the San Gabriel Valley's natural air inversion where smoke from burning grass lands and from their cooking fires settled low to the ground.

Then the Spaniards and Mexicans arrived to name canyons and mountains after figures in their religion: the Mexican named the mountains the Sierra Madres, Mother Mountains, which changed to the San Gabriels after their St. Gabriel, and 'Los Angeles' from the city of the Angels; the Verdugo Hills reminds us of the "robinhood' bandit Verdugo who roamed and robbed the area; the San Fernando and San Gabriel missions were named to honor St. Francis and the angel Gabriel, both special figures to the Catholic Spanish. Later on when the immigrants bought land from the vast Mexican land grants, the Ranchos, they brought names reminiscent of their former towns and cities along with names of special people: Glendale, Burbank, Indiana Colony (Pasadena), and Claremont.

Arriving in these new towns, along with those seeking health and a new life came the scammers, described by

Reverend Farnsworth, the stern looking goateed minister of the Methodist congregation in Long Beach in 1883, as "clairvoyants, palm readers, Hindu frauds, crazy cults, fake healers, Chinese doctors, and the like lavishly advertised.... here blossomed forth in truly tropical luxuriance." He continued with his invectives saying that "Vacant lots were crowded with spielers of all sorts—religious or political fanatics, vendors of cure-alls, of anything and everything to catch the idle dollar."[12] These were the 'street' people looking for the quick dollar. Additionally, a more focused group of men sought profit as their main draw, through land development by buying the big Ranchos, subdividing them into smaller lots, grading streets that remained unpaved and dusty, or muddy, and then promising water in their efforts to sell the new subdivisions to the people who built their homes and made their communities.

Failed land speculation, drought, and other factors quieted things down in the late nineteenth century until a tuberculosis epidemic in the eastern states caused much suffering to millions of people. Doctors at a loss for cures recommended warm, clean, dry air, and sunshine. Seeking health and reading of the Southern California sunny dry climate, the people came.

A Book of California Bunglows
Published by The Burck-Gwynn Company
1908

(from the Sherman Library)

This book was compiled by a design-build construction company that also provided financing and insurance. Essentially a complete package for the home of your dreams. Of interest is their use of river rocks (cobblestone) in parts of their bungalow homes. It was common for this company to use river rock for the fireplace, front porch walls, and piers for the roof. This was a modest home of 26 x 48 feet. "The porch of this home is what attracts attention. The heavy cobblestone balustrade...give proof of coolness on a summer day..." the builders advertised.

Developers complied with their desires for homes by buying large tracts of land for $1,000 per acre and then subdividing each acre into five or so lots, and selling each lot for one thousand dollars. In many locations the 'small farm' lot, with a 100' width and 300' depth were developed and sold to midwestern farmers who had made money selling their crops to the U.S. Government during WWI. These family lots served peoples' desire to raise their own food.

The simplicity and basic nature of early twentieth century technology may astound us today for it consisted of crystal radios with vacuum tubes, operator-controlled telephones and party lines, where a few households would share the same phone number, coolers with blocks of ice in a top compartment, recent electrifying of homes and businesses, unpaved roads, and cars that broke down regularly.

42nd Street east of Figueroa Street, a bungalow built in the early teens with an eastern touch–the sloping roof.

World War I put a stop to normal activities and stopped the migration to California. In those two years of U. S. involvement, 1917-1918, Angelinos built ships and airplanes, bought Thrift Stamps and Liberty Bonds, registered four thousand enemy aliens (Germans, of which my grandfather was included), banned the teaching of German in public schools, rationed gasoline, and voluntarily set aside days to do without meat, sugar, and wheat in an effort to help feed the American solders in Europe. Testifying to the Germanphopic reaction during WWI, the German-American Savings Bank changed its name to the Guaranty Trust and Savings Bank.

In the decade following WWI, while the rest of the nation experienced slow to no growth, Southern California's population grew plentiful with numerous subdivisions promoted, some even advertising lots with 'docking space' along the dry Los Angeles River. Growth,

along with movies, oil, oranges, auto and airplane manufacturing became major industries. And during these years, the native stones, having rested there for millions of years, became useful to a few people of independent thinking.

The settlers who wanted a cobblestone home hired a stone mason and helpers, the mason usually being an immigrant from countries where stone was used as the typical building material—like in Italy, southern Germany, Ireland, England, and Greece. The owner probably discussed with the mason the style of his future home, either a style familiar to the mason or the owner's preferred and familiar style. Most were not ornate, a few stone castles were built; some reminiscent of Irish or English stone castles. Others preferred aspects of the craftsman bungalow design popularized in England in the

Along Wilton Avenue just north of Wilshire Boulevard in the flat lands west of downtown Los Angeles, these homes show the typical residences built in the 1920s, so different than our stone homes.

1910s. In his book, *Biography of a City*, Caughey wrote, "Pervasive was the popularity of the bungalow (the word is from bengalese (sic) or...the California bungalow." [13]

It could be a modest house of 900 square feet or an elaborate and expansive mansion as seen in the Green and Green designed Gamble and Blacker houses in Pasadena. "It suited the land," Caughey writes in his 1977 book on Los Angeles, "and the life-style. It was preeminently a single family residence, called for a garden setting and that meant extensive lateral spread."[14]

Many of the stone homes were square or rectangular in plan, without roof overhangs, while others show that the owners came from the eastern states of snowy winters using steep sloping roofs with overhangs. Chimneys were built in all the homes for the fireplace was the source of heating until central heating became available in the 1920s.

While today we expect the basics such as unlimited

Montgomery Ward in their 1922 golden jubilee catalogue included the latest in women's and men's fashion for city dwellers.

electricity, natural gas, sewers, t.v., telephones, internet, kitchen appliances, and paved streets with sidewalks, the homes of the early twentieth century lacked all these luxuries. Some of these had not been invented, some were not available, some were too expensive. Vacuum cleaners were not invented then and cobblestone homes may have been chosen to keep out wind-blown dust from the dry dirt roads and open lands. Wood-sided buildings were drafty sieves; dust would blow through the many cracks between the boards.

Southern California's native shade trees, such as oaks and sycamores, were infrequent only to be found in canyons and washes; wood homes would get hot immediately in the summers. Life then required physical work. Maintained by their dreams, faith, and hope, the settlers found these homes provided practical shelter; they were places to sleep, cook, and be protected from the worst of the winds and rains.

People lived outside in the warming sunshine of Southern California doing their work of building new communities while growing their food. Inside, a few books, religious pictures and icons, along with a cherished and rare ancestral photograph, provided the typical decoration. Modern kitchen utensils would include a wood burning stove, wood kitchen counter and table, a tin sink, cast iron pots and tin pans, spoons, and knives. After

Dust and dirt and small creatures could easily enter the wood-sided building with its many cracks between boards. Drafty and uninsulated they required constant maintenance.

WWI, electricity became available and then appliances were invented, one of which was a rudimentary toaster whose sides flipped out to hold the slice of white bread, washing machines with hand cranks turning rolling tubes to wring out the laundry, radios whose frequent sound was static, and of course cars whose electrical spark plugs ignited the fuel to move the car and whose exhaust sparks ignited the hay spread on the dusty roads.

Before all this though there were the mountains, the tall granite mountains of Southern California. Culture is influenced by geography and its geologic shapes; Southern California became not only geographically defined by the mountains and the Pacific Ocean but also sprouted an indigenous vernacular of architecture during these years—the homes built of river rocks along the foothills.

On a forest path we may ignore the rock as a common item not noticing its texture, coloration, and subtle streaks of various minerals, unless you are a geologist or gold miner. I'm neither. Just a normal hiker who sees golden beauty in these granite stones. Study these for a while and an understanding emerges that in the hands of an artist these stones, bound together by Portland cement, could be elevated into works of art, a cathedral of stones. I wondered, "How are these stones made, where do they occur. Why?"

END NOTES

1 Carter, C.F., *Some By-Ways of California*, The Grafton Press, 1911, p. 152.

2 Ibid., p. 95.

3 The Resources Agency, "Survey of West 12th Street, Claremont," State of California.

4 Austin, M. H. *Land of Little Rain*, Houghton Mifflin, 1903.

5 Henstell, B. *Los Angeles: An Illustrated History*, Alfred Knopf, 1980, p. 13.

6 Caughey, J. and L.R., *Los Angeles, Biography of a City*, University of California Press, 1977, p. 211.

7 Weaver, J. D., Los Angeles: *The Erroneous Village*, Capra Press, 1980, p. 55.

8 Ibid., p. 67.

9 Bakker, G.A., *California History: A Topical Approach*, Wiley-Blackwell, 2002, p. 159.

10 Dvorak, J., *Earthquake Storms*, Pegasus, 2014, p. 10.

11 Campbell, P., *Survival Skills of Native California*, Gibbs-Smith, 1999, p. xv.

12 Farnsworth, R.W.C., *A Southern California Paradise*, self-published, 1883.

13 Caughey, J. and L.R., *Los Angeles, Biography of a City*, University of California Press, 1977, p. 224.

14 Ibid., p. 221.

CHAPTER 2

Granite Mountains

"...high rugged peaks rose on all sides...The higher mountains are of granite; outside are sandstones, lying in thick strata highly elevated, standing at an angle. Some of these strata rise in steep, bare rises, thrust up as it were 1,500 to 2,000 feet above the canyon, or 4,000 to 9,000 feet above the sea, forming grand but secular features in the landscape." [1]

William Brewer, 1862, a member of Whitney's geologic mapping team surveying California

The central portion of the San Gabriel Mountains in Southern California with its winter shroud of snow. That winter storm had to be cold, for the snow is low, at about the 4,000-foot level.

The story of the river rock homes begins three to four million years ago with a look into California's geologic history. You've probably been going along just fine not knowing much about geology. And that's okay if you'd like to skip this chapter for here we tell the story of the origin of the granite river rocks, how they became rounded into melon like shapes, and why the stones occur only in certain locations.

"The San Gabriel Mountains, known as the Sierra Madres until 1927," John McPhee writes, "are...young and rambunctious in geologic time, their slopes average sixty-five to seventy percent (very steep)—yet they contain the oldest rocks on the west coast of America." McPhee continues in his book *The Control of Nature: Los Angeles Against the Mountains*, "that every year seven tons of rock, soil and decomposing granite wash down off every acre of these mountains."[2] Yet each earthquake pushes up the mountains faster than they erode.

Cobblestones large and small, boulders rounded and rough, the result of millennia-old erosion. To many the stones have always been here, well at least for the four million years needed to form the range from which the stones came. They were present when earlier civilizations lived on Southern Californian's slopping foothill lands—the Chumash and the Tongva Indian nations, then the Spaniards and later the Mexicans at their rancheros. But the granite river rocks were not found everywhere in Southern California, neither at the beaches nor in the coastal ranges composed of sedimentary soils.

"Without the chaparral the mountains of the Southwest would be stark pinnacles and naked ridges," describes John Caughey, "the foothills barren, rocky slopes, and the valleys nothing but beds of cobblestones and gravel."[3] Thousands of times over thousands of years mountain canyons during torrential storms spewed gravel, plant debris, sand, and stones into the valley. These stones exist only in dry river beds and in the mud-filled alluvial fans formed at the mouth of these mountain canyons. An example of these torrential storms of earlier centuries occurred in the season of 1861-62 when fifty inches of rain inundated the San Bernardino mountains, foothills, and valley. [4]

McPhee wrote about geologist Leon Silver's thoughts on the San Gabriels, "It's a bedrock ridge up there. It's a weirdo wonderful block of rocks, the most complicated mountain range in North America. It includes the oldest rocks on the West Coast. The San Gabs look like a flake kicked around on plate boundaries for hundreds of millions of years."[5] The land at its foothills gradually rising, like one vast playa.

An alluvial fan at the base of Tujunga Canyon. Photo courtesy of Benjamin and Gladys Thomas Air Photo Archives, Spence Air Photo Collection.

McPhee continues describing the land as "A very long ramp, and remarkably consistent incline ended in the sheerness of the mountain wall. This broad uniform slope is where seven tons an acre (of soil, rock and debris from the steep walls) had emerged from the mountains, year upon year for a number of millions of years..."[6]

There are so many alluvial fans at the base of the mountain canyons that over these eons the mud flows have coalesced into this tilting plane of compacted soil and rounded rocks–the building material of our Rockitecture homes and churches. The entire south face of the San Gabriel is a granite block shattered by seismic forces, and thus what was once solid has been crushed into fragments with sharp edges to be smoothed and rounded by storms and erosion. These fans developed by the action of meandering rivers of mud and rock flowing from intense storms layered the land many of us live on. Sloping land barren of trees, but covered by sage, native grasses, with roaming mountain lions, deer, coyotes, and rabbits. The area was named by the various settlers Cresenta Valley, Montrose, Glendora, Cucamonga, Claremont, and Upland.

Later concrete flood channels would be built and isolated that million-year-old process of land building, preventing nature from doing her work.

Yet it created this land of granite stones of which Susan Suntree wrote poetically:

"2.5 billion years ago
Magma thrust deep in the earth's crust
Rises through the layers
So slow
Cools and rises
Shedding minerals as it rises
So slowly lightened
Crystallizing as it rises
To a quartz froth
Fermented to sparkling granite
Bowing the crust from beneath
Lifting land above the water
The first continents"[7]

The volcanoes and their magma that made granite were underground and erupted millions of years ago, and never reached Earth's surface. Over a long, long time, the magma cooled within this airless womb forming mammoth blocks of granite. Blocks the size of mountains. "It is a wonderful thing," geologist Keith Meldahl writes, "to walk on granite, a rock that forms five or more miles underground."[8]

Huge blocks of granite like Half-Dome in Yosemite, "formed underground as the convergence of two great tectonic plates generate huge quantities of magma."[9] What gargantuan forces raised these mammoth blocks to the surface which continue to rise to heights of ten and eleven thousand feet?

The San Gabriels and its two ranges. The San Andreas fault is at the top right under Big Pines and Wrightwood creating the smooth level valleys there.

Susan Suntree explains that,

"1.6 billion years ago
New bouts of magma
Rises and washed to the Western rim of North America
Planting the first layer of the coming mountains
San Gabriels, San Bernadinos,
Cooking the rocks already there into new minerals
To one day be named
The oldest rock in Southern California
A glittering layered confection: Mendenhall gneiss."
(whose common name is 'granite')[10]

Our California mountains rise up and surface by the forces of the slow shifting of the Pacific Plate as it moves northwards about two inches a year. In ten million years Los Angeles will sail by San Francisco, who is stuck to the North American plate. While at the same time the top surface of the Pacific Plate is peeled like a potato shaving as it squeezes under the giant North American Plate. But instead of scraps, mountains, valleys, and hills appear.

The tallest mountains in the west are being pushed up, some portions to a height of 11,000 feet. California is a state created by the movement of migrants from many lands, and was itself geologically formed by migrating land masses of two tectonic plates sliding past and slipping under one another occurring "at a rate slower than finger nails grow,"[11] geologist Lisa White tells us. She said this land "drew millions of...immigrants hoping to find their piece of the California dream. On this part of the world geology not only shaped the history, it is history."

The San Gabriel Mountains continue their growth today with each earthquake. (Today's map of seismic faults in California looks like a shattered mirror.) Over millennia torrential storms lasting over a month eroded the mountains to create the San Gabriel and Tujunga River washes, changing the mountain's face into gulches, canyons ridges and valleys seamed like an old woodsman's face, wrinkled and dried. And Time is not finished with its reshaping of the land.

During times of torrential storms, stepping into the swift rivers meant death to deer or to people; while in the

An automobile washed into a sinkhole after the 1934 flood in La Crescenta-Montrose. Courtesy of the Los Angeles Photographic Archive.

dry times, they could walk across these washes filled with river rocks without wetting their feet.

The San Gabriel range frames Southern California on the north side by its east-west orientation, which may have been a creation of seismic events from the north-west movement of the Pacific Plate and the south-west rotation of the American Plate. Like a playful boy moving his toy blocks, the Pacific Plate may have snagged the southern end of the range (when it formed and was aligned in the north-south direction) and spun the range counterclockwise into the east-west direction we find today. This playful spinning required a few million years. McPhee wrote that, "The Los Angeles Basin...has been squeezed about a centimeter (four-tenths of an inch, about the thickness of a pen) a year" for over two million years. "Laguna Beach (about 60 miles away) and Pasadena "...are fourteen miles closer together than they were 2.2 million years ago."[12]

"Here lies the magnificent reach of land
That will one day twirl ninety degrees as it's carried northwest
To stretch and sink, crumble and rise
Into plains, mountains, and valleys we will name
Point Conception, Cuyama Valley, Mount Pinos, Mount Frazier
Santa Ynez mountains and valley, Santa Barbara,
Los Angeles Basin, Santa Susanna,
Verdugo, San Gabriel Mountains, and Channel Islands"[13]

The Mother Mountains' growth spurts were 12 to 16 inches with each earthquake, yet she did not remain tall as the torrential rains eroded her growth. Shaped by these forces of earthquakes breaking and cracking and crushing the huge blocks of granite and the rain, biblical flood intense rain, washing them down the mountain sides tumbling and rolling and bouncing off each other in rivers and washes knocking off ridges and corners to smooth their uneven shapes into rounded boulders the size of trucks and basketball-sized river rocks.

The Spanish settlers named one of the rivers the San Gabriel River and until the mid-twentieth century it flowed year round, where fishing for trout must have been fun and its results delicious. Over the past century, those historic rivers and their floods have become an infrequent sight.

In 1876, Ludwig Salvator made an interesting but naive observation that "Earthquakes occur frequently, usually in August."[14] But, in the early morning of February 9, 1971, a clear and temperate morning, the San Fernando earthquake (known as the Sylmar earthquake) rose the San Gabriel Mountains eight feet in many areas with its magnitude of 6.6 strength.[15] And on January 17, 1994 in a cloudless morning of a comfortable sixty-five degrees at 4:30, without warning a 6.7 earthquake shook the land as far away as Las Vegas, rose the local mountains by twelve inches, and frightened Southern California residents for twenty seconds. This chaotic and destructive quake—neither the first nor will it be the last—collapsed buildings and bridges, broke gas pipes that resulted in fires, and damaged water mains hampering the dedicated efforts of our first responders. This event became known as the 'Northridge earthquake' resulting in billions of dollars of damage; yet the stone homes remained, some with slight damage others unfazed.

The citizens of La Cresenta needed to have a fallen river rock wall reconstructed in one of their parks. We've met Stuart Byles earlier and he relates this story: "After we finally convinced the city of L.A. to reconstruct the historic but failing retaining walls in our local park, we then saved a lot of local stones for their use." Byles said, "So, being a historic park to our community, we needed to use local stones. We saved the stones from a century old retaining wall that the local County Sheriffs department removed for more parking." Warming up to his story, Byles continued, "Well, we got a lot of volunteers, some trailers and put plywood sheets in the pickup beds, loaded 'em up and hauled all this salvaged stone down to the park for their mason. You see the color was right, it was local, it had that pinkish cast."

The stone mason hired by the city was not from the area and didn't know stone from one end of the San Gabriel Mountains to the other, like an auto mechanic who can't recognize the difference between a Ferrari and a Honda. Byles said, "He hauled-in grey stone, probably from the White Water quarry at the Mojave Desert in the east, dump trucks scattered this grey rock on the park site and he started building. We squawked to the city. Their mason had no idea what he was doing with that foreign stone.

"After some weeks, corrections were made and the local stones with the pinkish tint of the area were finally used to rebuild those century-old retaining walls." Byles smiled at their small but critical victory in keeping their stone heritage—a dedication to authentic local history and an artist's appreciation of the stones' beautiful textures, colors, and shapes, each unique much like the men of a century ago who built their individualistic and artistic stone homes. It could be that those who immigrated from Europe, England, Ireland, Italy, and Germany chose to build with river rocks not because the stones were there, but they probably desired the familiar 'castle' like appearance.

Beauty is important to me, and for the like minded look on. And for those who enjoy the historical and technical, read on. This will be a visual and written journey of art, history, technology, and geology in the search for California's stone homes.

But first...

A FEW GEOLOGIC UNDERSTANDINGS FOR THE CURIOUS

"The color of the mountains varies with each hour of the day...Now they are seen looming up before us, clear and distinct, each marking and furrow in their scarred sides as sharply defined as a knife-edge..."[16]

The table-sized granite boulders and basketball-sized river rocks, the stone for the stone homes, lie in the foothills and washes of the granite San Gabriel Mountain range, a sixty-mile-long and twenty-five-mile-wide transverse range with peaks soaring to 11,000 feet, peaks like Mt. Baldy, San Gorgonio, and San Jacinto. They are among the highest in the west. As we noted, they continue to rise with each earthquake; the San Gabriels range is a young mountain range growing like the teenager it is.

Geologists believe that the steep south front face of the San Gabriels attest to the young age of the formation and that "mountain building began about 4 million years ago, accelerated around 3 million years ago," and continues "rising faster than they are being eroded."[17] This growing range extends continuously from the shores of the Pacific Ocean to the sands of the Mojave Deserts.

The San Gabriel range is one of six within the United States that is oriented in the east-west direction. The other transverse ranges are all distant from each other, located in Alaska, Utah, Oklahoma, Alabama, and Arkansas. "Southern California's Transverse ranges" are so named as the book *Roadside Geology of Southern California* tells us, "because they are transverse to the general Northwest orientation of most mountain ranges in the western United States."[18] Recent understanding revealed that it "has been the ramming of Baja California into southeastern California" that resulted in Southern California "to be caught in a vice." Geologist John Dvorak said, "The area is being squeezed, and that has caused the crust to buckle" and raise up the San Gabriel Mountains; "an east-west line of mountains—the Transverse Ranges which run from Santa Barbara to San Bernadino..."[19]

Four centuries ago, in the land then named 'Alta California', the Spanish, looking for gold, rode up from Mexico and came upon this barrier naming the range 'the Sierra Madres', the 'Mother Mountains'; their massive size and extensive length impressing the early Spaniards. In prior centuries, the locals, the Chumash and Gabrielino tribes, named their Southern California home the 'land of smokes', as the natural temperature inversions of the air

1938 flood and mud covering Montrose continued building the alluvial fan.

sealed in the smoke from their activities.

Without the 'mother mountains' Southern California would never had occurred; it could have been a harsh land, like the Mojave Desert. For without the mountain, rain and snow would not occur, ground water would not accumulate, people would not settle in this 'land of sunshine', and Collis P. Huntington would not have mined for gold in the San Gabriels or his son may not have built the largest electric passenger rail transportation in Southern California, known as 'the red-car'.

Those immigrant residents were as yet to experience their new homeland's furious weather.

"It has now rained about seventy hours without cessation—for forty hours of that time…it has rained like the hardest thundershower…I never saw such rains before."[20] Though William H. Brewer wrote this in 1861 while surveying California with Mr. Whitney, similar storms would come in 1889, 1934, and 1938. Brewer noted that in December of 1889, "Sixty inches of precipitation fell at Kinneloa Ranch (northeast Pasadena in the foothills bordering on the small town of Sierra Madre) whose elevation is 1,500 feet and over 93 inches at Big Bear Valley at an elevation of 6,000 feet. Severe flooding occurred throughout Los Angeles with bridges destroyed and forty houses floating away."[21]

A 1963 California geologic study mentioned that "The (San Gabriel) mountains are almost exclusively a complex of crystalline rocks."[22] And in today's geologic terms, metamorphic and granitic rocks with areas of gneiss, schist, and igneous intrusions of other rocks containing minerals like pink or white feldspar, rust orange iron, and white veins of quartz.[23] This may seem complex to us, but just consider making a cake and combining all the ingredients, folding the flour into the eggs and adding other ingredients, maybe blueberries or chocolate chips, then baking the gooey mix in a hot oven watching it rise above the rim of the cake pan.

Just such a process occurred to the making of granite—except no one could watch it. Miles underground, molten magma cooled into granite and then tectonic movements pushed it through the surface forming our local mountains. While warm and pliable, seismic movements bent and distorted mammoth blocks of granite, like an industrial compactor crushing, bending, and deforming a car, creating the granite's streaks, freckles, and swirling patterns. For the San Gabriels, this

process may have taken just (a geological 'just') three to four million years.

But before this, the land mass had been moving eastward, taking its natural time and in no hurry, to rest on the shoulders of the North American continent. California was created by this slow accretion of land masses slowly moving eastward taking some hundred million years to add granitic land masses to the continent.[24]

This movement and the resulting seismic events made Southern California's landscape. As geologist John Dvorak describes in *Earthquake Storms*, "Almost every hill, every river, every canyon, and every fault has "...come into existence in the last four million years." To experience the stones, the canyons and the mountains without the need to lace up hiking boots, enjoy a pleasant weekend drive to the east side of Sierra Madre and up into the canyons above Baldwin Ave. to Mountain Ave. Turn right onto Sturtevant Drive and follow this century-old narrow road into a steep, dark rock walled canyon, Sierra Madre canyon. Here you will see and touch these exposed walls of rock, some 50 to 60 feet tall where "the outlines and color of every rock were preserved and the surface (of the wall) was reasonably flat."[25]

I arrived mid-day and the stones sat silent warming in the sun's intense light to be then cooled once again in the canyon's night breezes. Towards sunset the sun's oblique rays brought the stone wall to life dancing with sparkling crystals and shimmering mica, pink, grey, black, and white, all reflecting the day's changing light. This momentary brilliance returns every sunny morning and afternoon. Captivating, granite stones–solid yet vibrant. Their hard shapes come alive sparkling in the evening light.

The granite's former liquid condition appears in some stones as chocolate mousse ice cream swirls. The lazy 'S' patterns created, not by a spoon, but by the combined forces of heat, earth pressure, and the slow folding deformation of the warm rock being squished and bent under tectonic forces. Like a painter mixing his oil paints, earth mixed these minerals, and cooled them solidifying into granite rocks.

Geologist Keith Meldahl describes granite as "a handsome rock, crystalline and sparkling, light in color, like salt mixed with a bit of pepper and cinnamon."[26] Resting in these canyon washes and alluvial fans, a temporary stop in their multi-million year journey to the ocean to be recycled into the earth's crust, these flecked and streaked stones would add to the textured culture and history of California's indigenous river rock buildings.

The foothill lands continued to evolve, from extensive dry lands of chaparral to crowded lands covered with freeways, concrete storm channels that only engineers could love, and tasteless mini-malls, yet continues on its slow geologic journey of fires, floods, earthquakes, and drought all needed to form rounded granite boulders for

the creative to use and the fearful to avoid. This land of stones provided the basics to build many homes; some stones the size and weight of refrigerators and most the size of basketballs and melons with the same weight as a two year old child. There were homes of unique spirituality with soulful connection to the heartbeat of mother earth that no other material can provide. Now let's explore who built them?

END NOTES

1 Brewer, W., *Up and Down California, 1860-64*, ed. Francis P. Farquahar, University of California Press, 2003.
2 McPhee, J., *The Control of Nature: Los Angeles Against the Mountains*, Farrar, Straus, and Giroux, 1989.
3 Caughey, J. and R.A. Caughey, *Los Angeles: Biography of a City*, University of California Press, 1977.
4 Fisher, P.C., *The Mountaineers*, San Bernardino County Museum Association, 1972, p. 29.
5 McPhee, J., *The Control of Nature: Los Angeles Against the Mountains*, Farrar, Straus, and Giroux, 1989.
6 Durell, W. and G. Hise, *Land of Sunshine*, University of Pittsburgh Press, 2005, p. 192-203.
7 Suntree, S., *Sacred Sites: The Secret History of Southern California*, University of Nebraska Press, 2010, p. 30.
8 Meldahl, K., *Rough-Hewn Land: A Geologic Journey from California to the Rocky Mountains*, University of California Press, 2011, p. 53.
9 Farndon, J., *The Illustrated Guide to Rocks and Minerals*, Lorenz Books, 2018.
10 Suntree, S., *Sacred Sites: The Secret History of Southern California*, University of Nebraska Press, 2010, p. 30.
11 Nova Interview with Lisa White, episode 18, November 25, 2015.
12 McPhee, J., *Assembling California*, Farrar, Straus, and Giroux, 1993, p. 267.
13 Suntree, S., *Sacred Sites: The Secret History of Southern California*, University of Nebraska Press, 2010, p. 67.
14 Kipen, D., *Dear Los Angeles: The City in Diaries and Letters 1524 to 2018*, Modern Library, 2018, p. 266.
15 Sylvester, A.G. and E.O. Gans, *Roadside Geology of Southern California*, Mountain Press, 2016, p. 249.
16 Carter, C.F., *Some By-Ways of California*, The Grafton Press, 1911, p. 164.
17 Sylvester, A.G. and E.O. Gans, *Roadside Geology of Southern California*, Mountain Press, 2016, p. 249.

18 Ibid., p. 249.

19 Dvorak, J., *Earthquake Storms*, Pegasus Books, 2014,

20 Kipen, D., *Dear Los Angeles: The City in Diaries and Letters 1524 to 2018*, Modern Library, 2018, p. 13.

21 Bakker, G.M., *A California History: A Topical Approach*, Wiley-Blackwell, 2003, p. 289.

22 The Resource Agency. "Survey of West 12th Street, Claremont," State of California.

23 Sylvester, A.G. and E.O. Gans, *Roadside Geology of Southern California*, Mountain Press, 2016, p. 249.

24 McPhee, J., *Assembling California*, Farrar, Straus, and Giroux, 1993, p. 267.

25 True, J. M., *Painted Rocks*, Arts and Crafts Press, 1965, p. 132.

26 Meldahl, K., *Rough-Hewn Land: A Geologic Journey from California to the Rocky Mountains*, University of California Press, 2011, p. 49.

CHAPTER 3

The People Who Settled Southern California

"...intoxicating allure of American expansion, dreams of a better life, perhaps quick riches, cheap land and, most of all, a new start in a promising locale."[1]

"Within the past few years Southern California is (sic) become the Mecca of the constantly increasing number of winter tourists, as well as invalids and persons in delicate health who seek to avoid the...cold season of the northern states."[2]

"What struck me most," young John Regan wrote to his wife in Kansas City, "was the vast expanse of desert and the mountains. I thought they were beautiful. Right now the deserts are in full bloom and the colors are wonderful (lavender, blue, copper gold, red, yellow, green, white). I never dreamed there was so much color in the world. These mountains were a deep purple capped with snow which glistened high up in the sunlight, reminds one of a huge sundae."

The words and wonder of a young man raised in the flat states such as Kansas or Missouri, Regan's descriptive letter while riding the train picturesquely describes the untrammeled, pristine country of California—a heavenly wonder to one from the bland plains of the midwest or the industrialized smoky cities of the east. During the first part of the twentieth century this famous scene of beauty and bounty drew thousands of snow, smoke, and soot-bound citizens. "The stars in Southern California...seem nearer to the earth, have a warmer glow and lose all of the glitter of the eastern night sky. Here the stars seldom 'twinkle' (as the air in Southern California would have been clearer, free of soot and dust.)[3] "Shortly after 1880," the *LA Herald* stated, "Los Angeles began to attract a population of educated, refined, intelligent and moral people from the east...and gave a good deal of attention to the community of which they lived."[4] These important characteristics gave life to all the new southern California towns, and not one

Broken axles were common in these early car days.

became a ghost town, all the towns remained vital and expanded to their neighbors' town limits.

Though born in Murrieta, (west of Hemet and near the San Jacinto Mountains), in 1898, Irene Charnock's story exemplifies the stories of many of the immigrants.

Later this frontier child and her family moved to Pasadena and as a young woman she hiked the San Gabriel Mountains. And hiked and hiked and achieved the seldom granted '100 peaks' badge of the southern Sierra Club, a tribute to her trekking up one hundred peaks among the mountain manzanita and chaparral along animal trails and creating her own trails. Many of the peaks were 8,000 to 12,000 feet high where near the summits pine trees flourished in the rocky ground. These typical eight-to-ten-hour mountain hikes covered up to twenty miles and required stamina.

That is the common characteristic of most settlers to Southern California, the settlers who built the stone homes. With youthful strength they established new communities, changing the dry landscape of the valleys into thriving orchards of citrus and vineyards irrigated by miles of canals and hundreds of artesian wells where pressured ground water overflowed from shallow wells.

Irene Charnock, a typical early settler to Southern California

On her treks, in dry air and sun-heated hillsides, Irene Charnock must have scrambled over boulders and cobblestones in the deep cool canyons shaded by centuries-old California oaks, and probably wondered, How did these ankle-twisting boulders get here? In 1912 Walter Weyl, a writer for Time-Life books, described the culture of hiking during those early years: "During the second decade the old American habit of being in the great outdoors became known as 'recreation'...to be pushed rather like a religion."[5]

In 1910 the Charnock family settled in Pasadena. Two years later they drove a new invention, the car, on a long road trip to visit relatives. After a few days of dusty roads and sunny skies their slow driving probably caused them to stop in the young towns of Claremont, San Bernadino, Riverside, and Redlands where many of our cobblestone homes were built, close to the mountain along the foothills.

Maybe you've seen 1910-era autos with their big wood-spoked wheels and small noisy engines. Most roads then were unpaved, dusty, and rutted causing breakdowns and delays in traveling. Gerald Haigh remembers these roads for he grew up on them in early 1900s, "The roads were covered with straw to keep down the dust, each rancher

A weekend family outing touring the foothills of the San Gabriel Mountains in early 1900s.

covering part from the straw on his field. This worked fine until someone invented the automobile and they set the straw roads on fire..."[6]

Irene may have been accustomed to this new mode of travel and proud to have a father who could afford a car. Speaking in the enthusiastic way of a twelve-year-old girl, she wrote that "our family loved the out-of-doors and the beauty of everything."

Around 1900, the culture of hiking clubs, camping groups, and bicyclists expanded. They explored and tramped into the San Gabriels and other mountain ranges, many seeking these natural settings far from the chaos and sooty skies of industrialized cities. Responding to this return to nature movement was the handmade 'arts and crafts' style of homes and furniture which became popular along with the 'craftsman cottages' architecture which used cobblestones in to build chimneys and front porch pillars. Much of society throughout the U.S. favored naturalness and this preference may have influenced many immigrants to construct homes of cobblestones.

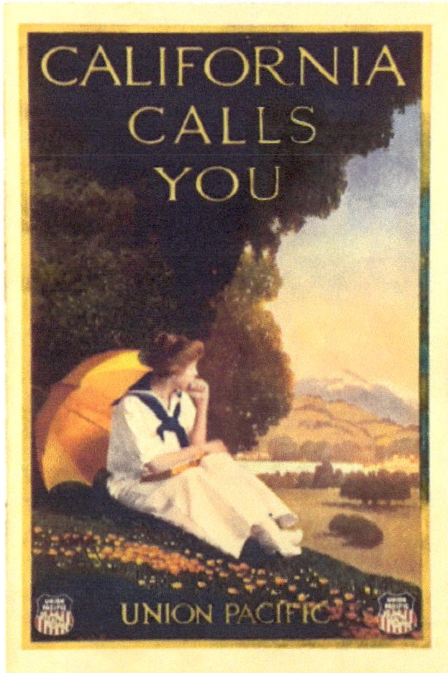

Above: ***Map of California, with a brief description of its Resources, Attractions, Topography, and Climate; Southern Pacific Company, 1915. Printed on the verso of the map is information about the state's farming resources.***
Above right: ***Railroad company advertising luring Midwesterners to California.***

To settle in undeveloped Southern California meant moving in with miles of chaparral-covered land, an occasional oak grove thriving on the ground moisture in canyons, rattlesnakes, coyotes, and rabbits. Along the mountain foothills streams would feed the bears, mountain lions, and deer living in the canyon shade of California Sycamore and Oak trees.

That photo of young Irene, who, judging by the fashion of her pre-WWI clothes, shown in her late teens, with petite delicate features. Though appearing refined, her accomplishments show her stalwart and rugged character. She taught children in Altadena and Pasadena schools as women had few choices of occupations then—either teach children, nurse the elderly and the wounded veterans from WWI, or help their father on the ranch until they married.

The young women and their life in this newly settled agrarian life "...were very good riders and had good saddle horses...rode with split skirts and Western saddles...they could shoot and rope as well as..."[7] any cowboy, Gerald Haigh recounted in a 1975 interview. During those years of sketchy medical skills, Irene lived into her eighty-sixth year, which was not unusual, for many settlers thrived in Southern California's clean air, year-long sunshine, and by eating locally grown fresh food.[8]

People settling in Pasadena and the other foothill communities emigrated from Italy, Mexico, Hawaii, Holland, Germany, Denmark, Texas, Kansas, Oklahoma, Illinois, Iowa, Missouri, and many other midwestern and eastern states. They brought with them their Protestant heritage, their manual skills, their mom's recipes, their culture, and their dreams. Also, many brought children.

California had been a desirable destination for people from all walks of life: "southern midwesterners were no strangers to California. By 1910, 103,241 Southwesterners, two-thirds of them Missourian, were already living in California. More than 300,000 others joined them during the big surge of the 1910s and 1920s. These settlers wrote home about their prosperity," encouraging family members to join them.[9] Most were tradesmen, farmers, or merchants and some were doctors. Many had limited education, the highest grades completed being the fourth to the seventh grade. They came with their families seeking freer open lands, and they came with their resilience.

Though the myth of the small farmer provided great promotional copy for the boosters, the sobering fact was that just nine large agribusinesses owned the vast majority of cotton production in California in the 1920s. And by the 1930s, four-fifths of agricultural production came from large agribusinesses, not from family farms.[10]

In *California of the South* (1888), the authors wrote: "The health seeker who, after suffering in both mind and body, after vainly trying the cold climate of Minnesota and the warm of climate of Florida...and noting that he is losing ounce upon ounce of flesh, that his cheeks have sunken, his appetite capricious, his breath more hurried.... turns a gleam of hope toward the Occident,"[11] by which they meant Southern California.

Arriving by train they traveled through the Rocky Mountains and the great deserts of the southwest experiencing the cold nights in the mountain climate. They warmed their meals on the one coal stove in the Pullman passenger car and spent hot days in the deserts opening the slide down wood-framed windows for the slight cooling from warm breezes. Settlers arrived

Los Angeles train depot in 1899

by the thousands when travel time from the eastern states dropped from four months to four days. The completion of the Union Pacific Railway after the Civil War in 1869 (the transcontinental railroad) provided a fast journey compared to the original settlers' journeys with wagons and oxen and walking three thousand miles.

In 1870 an 'Emigrant class' train ticket for the long journey cost $60. That would allow the passenger to travel thousands of miles on wooden seats in a wooden Pullman railcar. Southern Pacific railroad connected Los Angeles to Kansas and soon after price wars decreased fares to $1.00. That savings would have been absorbed by the rising prices of land that Southern Pacific Railway owned in Southern California and sold parcels to their passengers.

Los Angeles Boys Home 1890

A short pamphlet published by the San Bernardino County Museum in 1972 included many interviews of the relatives of the early settlers and this brief statement gives an accurate idea of their life. "The homesteaders mostly raised chickens, rabbits, apples, pears, cherries, peaches and grapes. They dug their wells by hand and cemented them in. These wells were about ninety feet deep and some of them are still in use."[13] In 1918 the Pasadena City Directory included a listing for the 'Chinese Methodist Episcopal Mission'. Chinese culture in Pasadena did not seem right with its Anglo-Saxon population. Yet after laboring and building these railroads, many Chinese people migrated to Southern California in search of gold, and not finding much, later dug and blasted the 7,000-foot tunnel through the San Gabriel Mountains for the Southern Pacific Railroad Co. After this, for one dollar per day, the Chinese laborers blasted and dug more tunnels to build the four hundred mile aqueduct stretching from Owens Valley to bring water to Los Angeles in 1913.

Another listed occupation in that Pasadena Directory of the early 1900s tells of the westward migration of freed slaves after the Civil War, people seeking a new life and fleeing the southern states Jim Crow laws. These people's occupation was listed as 'bootblacks', those who polished

Pittsburg—1910

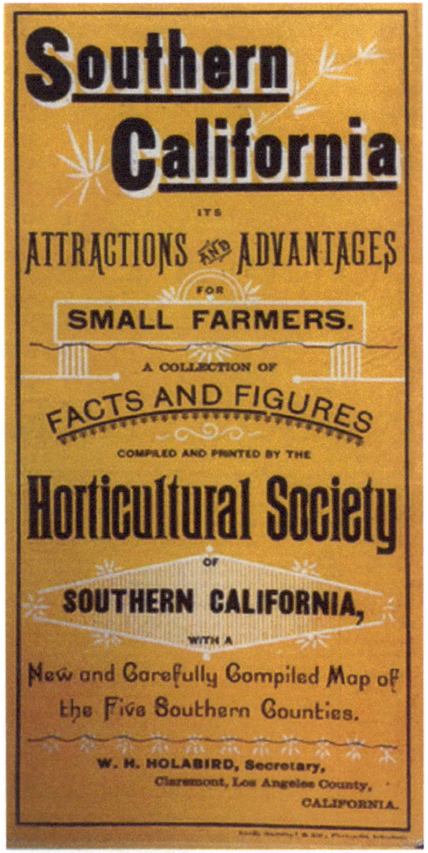

away dust and mud from walking on unpaved roads in city shoes.

People of modest income lived in Pasadena and Altadena, towns then known for the homes of wealthy retirees and winter mansions of vacationing eastern capitalists. It appears that living costs then could accommodate all levels of income as Kate Sanborn, visiting from New Hampshire, in 1906 mentioned in her novella describing her travels in Southern California, "Pasadena has no ragged shabby outskirts; the poorer classes seem to be able to own or rent pretty little homes...all well kept and attractive."[14]

The immigrants lived during mass changes in the U. S., changes to their social lives, economic lives, and working lives. These were the years of no taxes, opulent wealth to a few, and the seeking of utopian communities by many. These were the years of increasing women's influence during the suffrage movements and obtaining their right to vote. Also deep recessions occurred in the 1860s and then again in 1907, and, with industrialization, people from the farmlands moved to city tenements to labor in unsafe factories. Those fortunate enough to leave the industrial servitude of factory life did so. The railroad companies spread abundant advertising while lauditory letters from relatives living in SoCal encouraged "a steady flow of Easterners eagerly sought a new land of milk and honey."[15]

Traveling west by car or by train, they passed the Rockies and for the first time "saw something on the horizon other than the horizon." People arrived and began a new and different life. "A country conceived and populated by wanderers," William Least Heat Moon tells us. These settlers were "migrants, immigrants, voyagers, vagabonds, most of them believing in the far side of the rainbow, optimists for whom a road is an enticement beyond resistance."[16] Add to these the people affected by the tuberculosis epidemic desperately seeking relief when their doctors could not help them and sent them west, and the veterans of WWI who inhaled mustard gas damaging, their lungs and sought clear dry air in desperate hopes of healing. Others sought the nostalgia of simpler times as evoked by the songs of the day "In the shade of the Old Apple Tree" and "By the Light of the Silvery Moon." And Europeans wanted to escape WWI's carnage and great destructions of their cities and towns. Many left the city life of Los Angeles to settle in the sloping lands of the mountain foothills and build their stone homes.

"During the early 1900s, more than seven of every ten immigrants came to America from (the) central and southern European countries."[17] And they brought with them their manual skills learned in their homeland such as the Germans excelling at carpentry and the Italians masons mastery of stone construction. "Thousands of southern Italian the immigrants who knew masonry. By 1890, two out of every three masons...were immigrants."[18] They also brought, as we all do, with them their culture and habits. From generations of tradition, many built in cobblestones, as the Irish, northern Italians, Greeks, and the mountain living Germans thought this was the normal way to build a solid home, an ancient psychic hunger for stability and firmness.

Throughout the first three decades of the twentieth century, of the people living in Los Angeles 20% were born there, 60% came from other states, and 20% were foreign born. The largest growth decade in Southern California occurred in the 1920s, beginning in1920 with 936,455 residents burgeoning at the end of the decade to 2,208,492.[19] Those who migrated from eastern cities in the U.S. left apartment buildings seeking the opportunity to tend their garden of young fruit trees, seasonal crops, flowers, and herbs in clear warm skies.

Catalina Island harbor around 1910

The plein-air painters of that time attested to the desire for natural settings and nature's materials.

WHY SOUTHERN CALIFORNIA IS PROSPEROUS.

SPENCER G. MILLARD, Republican Nominee for Lieut. Gov.

SOUTHERN CALIFORNIA is justly celebrated for its superb climate and its perpetual sunshine; for its invigorating ocean breezes, and its beautiful scenery of mountain and valley; for its abundant water supply, and the fertility of its soil.

It is now becoming celebrated, also, for its prosperous cities, its thriving villages, its orchards and its vineyards, its seaside resorts, and its villa homes.

The building of the first transcontinental railway into Southern California awoke the city of Los Angeles from a sleepy adobe village. Its growth from that time was steady and continuous.

The building of the second transcontinental line into Southern California precipitated the boom of 1886–7. Then came the influx of Eastern people with their money, their push, their enterprise and their intelligence. The combination of these with the intelligence and the natural advantages already here transformed Los Angeles from a Spanish town into a beautiful, prosperous and American city; built up other cities and towns all over Southern California, and transformed great ranches of dry pasture land containing thousands of acres into ten, twenty and forty-acre tracts of irrigated land, dotted all over with orange and lemon groves, vineyards and deciduous orchards.

Promotional articles and ads filled magazines like *Home and Garden* and *Sunset* with articles and photos of flowers blooming in winter and displays of flowers in town parades as seen in the now famous annual Rose Parade of New Years Day that was inaugurated by Pasadena's Valley Hunt Club on January 1, 1890.

In the beginning of the twentieth century in Southern California, the settlers depended upon the cash crops of citrus and vineyards. Land speculators, the 'real estate people,' hustled and promoted then, as now. Visionary people founded communities, schools, colleges, and city governments. Hiram Reed, in 1895, mentioned the settlers "were that class of people who regard public schools not as a mere ornamental appendage, but as one of the prime necessities of a civilized community."[20] Schools and churches provided the central focus of spiritual and social activities.

Everything was imported—the people, plants, animals, wood, Portland cement, and water—except for the native building materials: cobblestones, sand, gravel, and gypsum.

Tourists traveled to Southern California just to experience the unbelievably famous sunny climate. Many stayed or returned to buy the climate of sun-filled skies. These people were filled with the "natural human desire for autonomy and independence"[21]

Practical, consistent with their nature, intent on staying, quite a few built their homes with cobblestones. Not looking for the hit-n-run buck, these settlers were making new lives and planned on staying in their new solid homes. And like most of our creations, these cobblestone buildings "were the product of a very particular combination of circumstances, unlikely to repeat themselves."[22]

The industrialization of the 1900s spawned revolts to the constant exploitation, materialism, and contamination affecting peoples' lives. Not much has improved in the last one hundred years for then as now "America is in a period of clamor," Walter Weyl wrote in 1912, "bewilderment of an almost tremulous unrest."[23] Thousands saw relief in the orange treed dream of Southern California, "the west became a place of hope and opportunity."[24] Bill Bryson describes the moment when having "emerged into California, into warmth and sunshine....and the delicious odd feeling of warm sunlight on your face and bare arms in December, a winter without winter." But today the disaffected do not have new open lands on which to settle and start new lives.

Yet, we must understand that not all people exhibited strong characters. In the early 1900s and after WWI, Sam Waters in his book *Houses of Los Angeles* described how "...a repository of wild cat entrepreneurs (seeking oil), hair-brained politicians (many corrupt), religious quacks fleecing true believers (promoting communes) and ...Hollywood producers; in other words the dregs of society"[25] roamed Southern California.

While men of independent means built their mansions in Pasadena, Altadena, and southwest Los Angeles next to the new University of Southern California, men of independent nature built river rock homes in the foothills of the San Gabriels. Here winds blew at 40-50 miles per hour, toppling over barns, garages, and filling homes with dust and insects. Torrential rains washed wood buildings down the sloping land while cobblestone homes remained unaffected by either dust or mud.

The big wave of settlers came to Southern California after the depression of the 1890s. But later the hardships of rationing during WWI, which ended in 1918, and the 1917-18 Spanish influenza, which infected millions, killing 500,000 people in the U.S. alone, hindered resettling efforts.

In the 1920s people bounced back from those years of death, wars, and disease unleashing the carefree energy of the Roaring Twenties. During that decade funds became available and also during that decade most of the stone homes and churches were built.

Describing his view of those settlers, Franklin Walker in 1950 wrote, "For in spite of the eagerness" of real estate land developers and promoters, "to capitalize on a Spanish past...the people were essentially homely in their tastes;" and he reinforces our understanding that "the domestic interests were basically as those of the East and Middle West, the regions which supplied the bulk of the new settlers."[26]

In those new foothill towns, such as Sunland, Tujunga, and Claremont, the scarcity of doctors and

In the early 1900s articles throughout newspapers and magazines proclaimed the wonders of Southern California.[12]

This historical photograph of around 1890 taken in the winter shows Tujunga town built on an alluvial fan at the base of the San Gabriel Mountains and Big Tujunga Canyon. Notice the low snow level then.

hospitals required that many learn the skills of healing, such as treating colds, broken bones, and cuts. "There were mustard plasters for coughs...and the hot mustard foot bath...for a runny nose." Though sounding simple in their methods these homey remedies may not have worked as desired for frequent infant mortality occurred leading to the need for large families. "In a day of high infant mortality, there was always the dread of diphtheria, (which) brought death to hundreds of Southern California children."[27]

Nikola Tesla's and Thomas Edison's inventions brought electricity to the homes and offices. In some rural areas of SoCal. though, it was irregular for their local hydro-electric generators depended on the sporadic rainfall and river flows. Horse drawn wagons and Model T trucks, the size of horse drawn wagons, provided transportation for building materials and store products. Local farmers markets gave households an outlet for selling excess garden vegetables, chickens, eggs, and summer fruits. Hunting and fishing provided occasional protein, but not fat as these animals lived off the canyon lands climbing hills and chasing their food. These were the original 'free range' sources of meat.

Julius Rakasits bought a plot of land on Cross Street, where, with the help of his family, he built a house out of stone that took 20 years to complete. In this late 1920s photograph, Rakasits and his kids are working on the house. They hauled much of the stone from Tujunga Canyon because Rakasits felt that the stone was "cleaner"—true rounded river rock. The house still stands. (Courtesy of Joe and Linda Rakasits)

The typical worries of the day may have included protecting your gardens from becoming a banquet for night time critters to mending a wind damaged home and caring for both family and community, and building those communities.

Most emigrants and settlers had adequate means to buy land and build new homes, the frugality learned

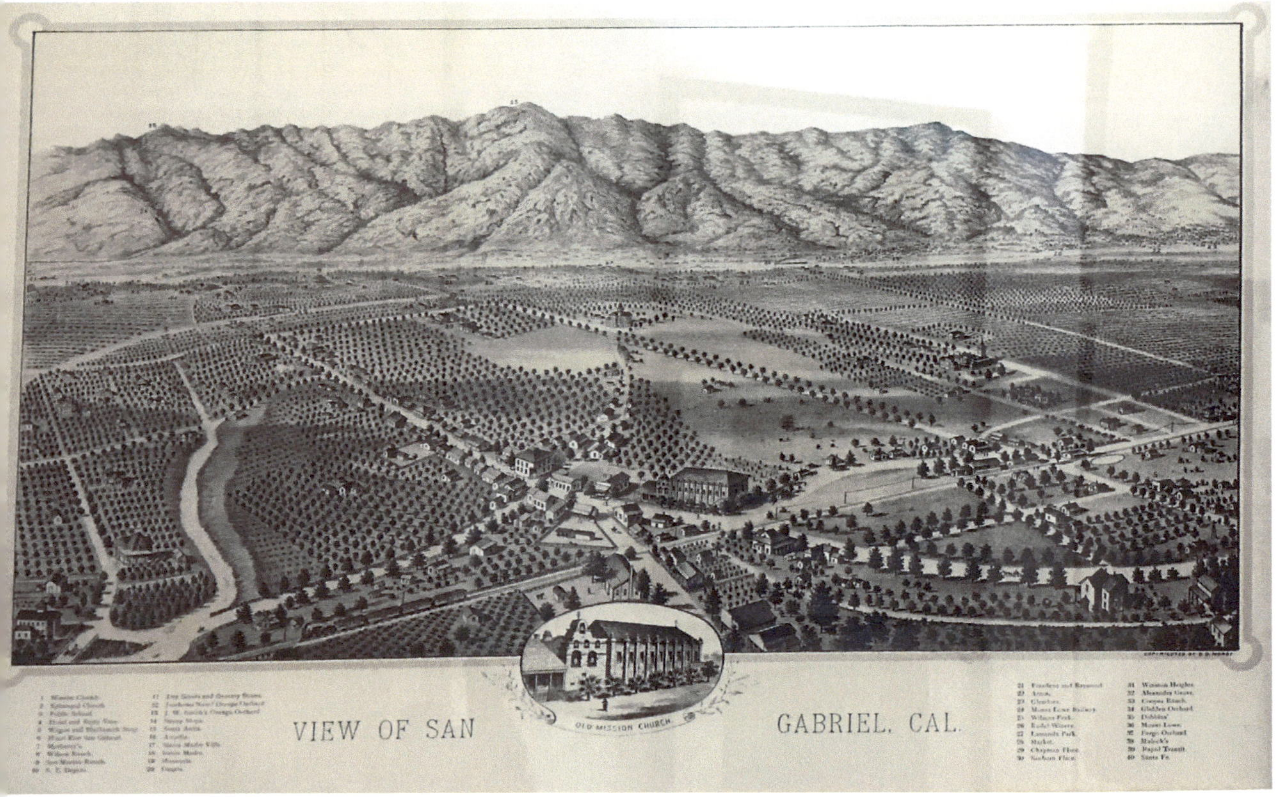

I found this rare print at the San Gabriel Mission, which I hope survived the recent and sad fire. This drawing by D.D. Moore illustrates San Gabriel Valley shortly after 1893. Pasadena is at the upper left and the Mt. Lowe Railway is noted above it on the upper left mountain peak.

during tougher times prevailed in most decisions. In 1913 $1 would buy the equivalent of $26.30 of goods and services in 2020. An acre of bare land may have cost $100, but the land came without services, without electricity or roads in some cases. Police and fire departments depended on volunteers. When telephone became available they were placed in select locations, like the community hall, fire station, or well-off person's home. Sewers did not exist but septic tanks did and were connected to the drain pipes of the recent technology of piping and indoor plumbing.

Family entertainment "had no difficulty having a good time. More self-sufficient than the present day family," wrote Walker in 1950, "it even provide(d) its own music with fiddles and other instruments."[28]

A settler's story describes life a century ago. Mr. Trout tells us,"When he (his brother) heard we were goin' to California he decided to go along...We drove out over the Santa Fe Trail." Traveling through the plains and then the desert of New Mexico, they probably stayed a night or two in Santa Fe or Albuquerque admiring and wondering at the silver jewelry of turquoise and silver of the Navajos while repairing their car.

Mr. Trout held many different jobs—self-employed candy maker, selling furniture polish from office to office, which paid him $35 a week; his girl friend worked as a beautician. Later he would become a"contract man" for Prudential Insurance, which paid about $25 a week, this varying as he sold on commission. Many others in the mideast may have been tenant farmers and looked to do better.

For those who settled in the foothills, they were comfortable with the rural setting where lives were based on the time of day and the seasons, not the time clock in an office or factory. Living their days outdoors in the sun and dust, heat and cold, their home would be considered special and critical: giving protection and shelter to relax from one's daily labors, a place for families to gather in the parlor to read, play games or play music during the evening. "Certain powers in the psyche (are) common to all mankind," Joseph C. Strong mentions, and walls of granite may have responded to some of these psychic needs. "California once supported North America's largest aboriginal population: a combination of favorable geography, climate, and abundant nature attracted them.... it was home to ten percent of the continent's people... archeological sites in California attest to that the search for a better life in a congenial climate is no new idea."[29]

Now in the late nineteenth and early twentieth century people "weary of war, strikes, Red Scares, and politics of any kind"[30] migrated from upstate New York and Pennsylvania, where they had built and lived in stone homes and worked in stone barns, brought with them this familiarity and constructed stone buildings. Now in the new age of automobiles the stone barn turned into a stone garage. It could be understood that many migrants wished to repeat their traditional way of life, their roots. "The people were essentially homely in their tastes. Domestic interests were basically as those of the East and Middle West, the regions which had supplied the bulk of the new settlers."[31] And those who chose to live in the Southern California foothills rather than the crowded central city of Los Angeles tell us much of their character and independent natures.

"By 1906 Southern California was in the midst of her third boom which brought more tourists and settlers" subdividing large land tracts, the former Mexican ranchos of thousands of acres, grading roads, installing water piping and expansion occurred throughout the valleys until 1917 when "doughboys were training in the California camps to go overseas (WWI)...and the era ended."[32]

Up to that time and to the end of 1920 when a national census occurred, population grew from "227,052 in 1890 to 337,328 in 1900, to 808,445 in 1910, and reached 1,423,756 by 1920,"[33] Over those thirty years the newcomers arrived in an average of 111 per day. The largest growth occurring between 1910 and 1920 of over 600,000 in that decade. And in that second decade of the twentieth century the aircraft industry expanded and in 1910 held its first western air show at Dominguez Airfield while ten years later Douglas Aircraft began production in Santa Monica, and the motion picture industry moved from New York to Hollywood and Culver City, and with both vision and chicanery the Owens Valley aqueduct brought water 300 miles to Los Angeles by 1913.

During those three decades, Henry Huntington built and consolidated local rail lines to create a trolley rail line known as the 'redcar,' named after the red painted wooden passenger train cars, providing passenger rail access throughout Southern California "connecting Los Angeles north as far as the city of San Fernando, east as far as Riverside, south as far as Newport Beach, and west to the ocean...It transported commuters who wanted to sleep in the woods and have an office in the city."[34] One of these towns 'out in the woods' began its life on the foothills above Pasadena. The railway served Pasadena but did not extend up the hill to the adjacent town of Altadena. Before cars, you'd ride in a horse drawn carriage up the steep slopes to your home overlooking the large tracts of orange groves in the San Gabriel Valley.

Our journey begins there, then we travel east through the foothill towns of Sierra Madre, Glendora, Claremont, and Upland, and to cities beyond in search of our cobblestone homes.

END NOTES

1 Woolsey, R.C., *Migrants West*, Grizzly Bear Publishing, 1996, p. 29.

2 Carter, C.F., *Some By-Ways of California*, The Grafton Press, 1911, p. 152.

3 Ibid., p. 25.

4 *Los Angeles Herald*, March 6, 1910, UCR Center for Bibliographical Studies and Research.

5 Weyl, W., "The New Democracy," *This Fab Century, 1910 to 1920*, Time-Life Books, 1969, p. 102.

6 Haigh, G., *Straw Roads: A Story of Simi Valley*, Simi Valley Historical Society, 1975, p. 77.

7 Ibid., p. 77.

8 Johnson, K.J., "What Were Lake Elsinore and Murrieta Like in 1910?" *Press Enterprise*, February 28, 2019.

9 Henstall, B., *Los Angeles: An Illustrated History*, Alfred Knopf, 1980, p. 13.

10 Bakken, G.M., *California History: A Topical Approach*, Wiley-Blackwell, 2003, p. 159.

11 Woolsey, R.C., *Migrants West*, Grizzly Bear Publishing, 1996, p. 29.

12 Bowen, E.B., *Annal of Early Sierra Madre*, Sierra Madre Historical Society, 1950, p. 107.

13 Fisher, P.C., *The Mountaineers*, San Bernardino County Museum Association, 1972, p. 21.

14 Sanborn, K., *A Truthful Woman in Southern California*, Gutenberg.org, 1906.

15 Walker, F., *A Literary History of Southern California*, University of California Press, 1950, p. 102.

16 Moon, W.L.H., *Writing Blue Highways: The Story of How a Book Happened*, University of Missouri, 2014, p. 44.

17 Bayre, M.A., *The Influence of WWI on Residential Architecture of Pasadena*, Thesis, University of Southern California, 1989, p. 94.

18 Ibid., p. 94.

19 *Los Angeles Herald*, March 6, 1910, UCR Center for Bibliographical Studies and Research.

20 Reed, H., *History of Pasadena*, Pasadena Historical Company, 1895, p. 168.

21 Judt, T., *Ill Fares the Land*, Penguin Books, 2011, p. 152.

22 Ibid., p. 152.

23 Weyl, W., "The New Democracy," *This Fab Century, 1910 to 1920*, Time-Life Books, 1969, p. 102.

24 Bakken, G.M., *Encyclopedia of Immigration and Migration to the American West*, Sage, 2006.

25 Waters, S., *Houses of Los Angeles*, Acanthus Press, 2007, p. 11.

26 Walker, F., *A Literary History of Southern California*, University of California Press, 1950, p. 151-2.

27 Ibid., p. 154.

28 Ibid., p. 78-9.

29 Zack, M., "Where Sierra Madre Came From" in *Southern California Story: Seeking the Better Life in Sierra Madre*, Sierra Madre Historical Preservation, 2009.

30 Evans, H., *Do I Make Myself Clear?*, Little Brown Company, 2008, p. 186.

31 Walker, F., A Literary History of Southern California, University of California Press, 1950, p. 151.

32 Ibid., p. 227.

33 Ibid., p. 227.

34 Ibid., p. 231.

CHAPTER 4

Altadena, Arroyo Seco, and Some Technical Thoughts

"That kind of work belonged to the past now. It ended with the Depression and seemed unlikely ever to be done again."

Wendell Berry, *The Dark Country*

My and Scott's journey sends us eastward along Foothill Boulevard, the four lane highway that follows the base and almost the entire length of the San Gabriel Mountains from San Fernando Valley to Rancho Cucamonga. A fifty-seven mile expanse of road now bordered by fast food places and strip malls of indescribable plainness. But during the late 1800s and into the early twentieth century in Cresenta Valley, Foothill Boulevard was named Michigan Avenue, possibly a nostalgic reminder of the early settlers' home state. Michigan Avenue was a dusty wagon road and the only connection for tourists and commerce between Cresenta Valley, Pasadena, and Altadena. Yet a wide canyon must be crossed, the famous Arroyo Seco Canyon separates the communities. It is a historical canyon which in 1896 Hiram A. Reed noted in his seven hundred page encyclopedic book, *History of Pasadena*, that "Arroyo Seco is Spanish for 'dry creek,' and this great branch of the Los Angeles river has been called that name almost ever since Los Angeles was founded in its pueblo days of 1781."[1] Though this 'creek' did go dry in the summer and autumn seasons, it becomes a wild torrential river during heavy rains. In the 1930s the Army Corps of Engineers built a large concrete channel to control the flood waters.

ALTADENA

Driving east through old town Pasadena on Colorado Boulevard we arrived at Lake Street in search of a historic stone community center in Altadena. Turning north, Lake Street extends in a straight line ascending up the sloping foothills. Surveyors prefer straight lines and this street was a surveyor's dream. Lake Street climbs from level land in Pasadena and steepens, steepens more and more into the foothills and into the community of Altadena, finally ending at the base of the sheer rock face of the San Gabriels where the street can go no further. As we drove I realized that all this soil at the foot of these mountains was once the mountain itself shedding its skin of chaparral,

The cover of the book by Michele Zack

mud, and rocks with each torrential storm. A few miles further we saw the building, at the very top of Lake Street.

We arrived, parked the car, and without exiting I stared out the window. A few yards away stood a masterpiece. Its stones glistening and sparkling, swirling like a Van Gogh painting. Textured, thick, stout Davies Center anchored its street, Mountain Curve Avenue, appropriately named for it follows the base of the mountain. Leaving the car we looked south, down the sloping foothill and beyond where past the building-covered Los Angeles valley far below we saw the Pacific Ocean, thirty miles away. It was a clear morning.

WILLIAM DAVIES COMMUNITY CENTER 568 EAST MOUNTAIN CURVE AVENUE

Mr. Farnsworth donated the land to the new community, possibly because the land was unusable for groves so high up the mountainside "where only wild animals and wild vegetation lived."[2] Upon finding yet another stone masterpiece, we became transfixed and impressed by the sheer energy and skill of the masons for their creations show us undeniably their caring craftsmanship. The buildings speak of the new community's belief of the importance of community life. And that is a quiet and salient message of the rock buildings.

Davies Community Center is a large craftsman rock and wood shingle building that overlooks the San Gabriel Valley, as it has for about ninety years, untouched by decades of earthquakes, smog, wind, fire, and floods. The Santa Anas, those dry hot winds, race through these mountains and foothills sucking the little moisture remaining in the land. A spark from lightning, electrical wires, or a careless or clueless smoker could ignite the dry creosote brush and grass, causing fires that could take on unrestrained lives of their own. These forces of nature have destroyed many homes and buildings along the south edge of the Sierra Madres, but I'm grateful many have remained including this survivor; it remains the pride of Altadena being in continual use for community performances, dances, and classes.

The Depression years of the 1930s in the U.S. brought massive unemployment, leaving millions in economic despair. Franklin Roosevelt, during his third presidential term, became determined to do something about this human suffering and created the Works

The Davies Community Center, 568 East Mount Curve. Photo taken shortly after its completion. Notice that the street was dirt and the small size of the foliage.

The street view today.

Progress Administration. Millions of destitute but willing workers were employed and created thousands of public facilities such as roads, dams, the Hollywood Bowl, and the William D. Davies community building. "The beautiful two story stone community building was completed in 1934," states the plasticized Charles S. Farnsworth Park History article which hung on the wall in the basement office of the Park Ranger, who showed us around. Women's groups were most popular then and the interior was outfitted with funds raised by Altadena Women's Club and Altadena Women's Circle. Notably, the article reminds us its building shows "the careful construction of local artisans."[3]

The afternoon of our visit an Irish dance class was scheduled in the large first floor room with its massive stone fireplace.

We thanked the congenial Farnsworth Park Ranger, who was rightfully proud of his responsibility, for his tour and drove down Lake Street, past a couple magnificent early century renaissance churches and wood-framed homes of various styles, to the Altadena Historical Society which is located on what else but Altadena Drive. The friendly docents brought out books on the early nineteenth century providing information on the details of life.

FOOD

Under the pleasant guidance of Jacquie Lawson, we reviewed local papers of the early twentieth century and found this delicious 1935 recipe, which was probably the same recipe used by contributor Nellie Maxwell's mother, her grandmother and, maybe, even her great grandmother, and no one updated the spelling of 'sirup'.

Life didn't change much then, except when you left

Historic photo taken shortly after construction and before the amphitheater at its base.

your family home and roots to move to California. And move they did, fast and for reasons—some to hide, some to retire, some to make a fortune, and some to heal.

LA CASITA SANITARIUM

"'When she was eight years old, she told me about her first vivid memory,' her grandmother told her," Lara Dotson-Retna wrote in her November 2019 article for *The Washington Post*. "'She was a toddler, and her mother, my grandmother, was dying of tuberculous.' She remembered being kept away from her bedroom and being pushed back as she tried to edge closer...This recollection of (my grandmother's) her mother on a bed, her arm extended toward her, is her worst memory...as it is the only crisp memory she has of her mama."[4] This sad story represents millions of peoples' suffering during the decades of tuberculosis epidemic in the late nineteenth and early twentieth centuries.

Dozens of sanitariums were built in the San Gabriel Mountain foothills and canyons to accommodate the influx of thousands of 'tuberculors' to the sunny climate of Southern California. As a child in the 1940s and 1950s, I was warned not to go to the Sierra Madre Canyon. I didn't know why, but it was a mysterious place. Later I discovered its other name, 'tuberculor' canyon. Many with tuberculosis settled there seeking the dry warm air and also because they were banished to the remote area, as tuberculosis was contagious. Today this area is a serene tree-shrouded canyon of gurgling water and Mary's Market, then a local grocery store and today a local cafe.

In the book *Altadena*, I found a picture of the former La Casita Sanitarium built to help tuberculors. Built of river rock in 1887 for Dr. Adele Gleason it was situated on a parcel near the Arroyo Seco bluffs, on the east side of Altadena and the source of the river rocks. The sanitarium was demolished probably after the epidemic scare had settled and people may have thought the building contaminated. Looking closely at the photo you may

KITCHEN CUPBOARD

By NELLIE MAXWELL

We may live without painters
Or writers or mummers,
But civilized man cannot
Live without plumbers.
—The Contributor's Club.

Good Things to Eat

THE following recipes, gathered from various sources, are all favorites of some one:

Harding Waffles.—Beat the yolks of two eggs and two tablespoonfuls of sugar together, add a pint of milk and a pint of flour alternately, sifting the flour with four teaspoonfuls of baking powder; add a teaspoonful of salt, two tablespoonfuls of melted butter and the stiff beaten whites of the eggs. Bake on a hot waffle iron and serve with maple sirup.

La Casita, "Little House" Sanitarium in Altadena, now demolished. Notice the liberal use of river rocks in its construction.

see the land slope.

The cause of tuberculous was unknown at the time but doctors prescribed dry, clean air, and sunshine. Many emigrants to Southern California swore by it. Eastern industrialized cities of those days of the early 1900s resembled the smog-filled days of the 1960s in Southern California and Denver.

A tuberculor herself, Josephine True cured her ailment in Cathedral City during the 1920s and after wrote that she noticed, "Children with bronchitis or tubercular trouble thrive in the dry sunshine...the climate is a blessing and great curing agent for all ages."[5]

COLONEL GREEN'S GATE HOUSE

Close by stood Colonel Green's wood-framed mansion, now demolished for the Altadena Library. But Green's stone gate house remains. Today the library parking lot is the best public vantage point to see the stone gate house. After Green made a fortune selling 'patent' medicines like Green's August Flower, he built the mansion in 1888 as a 'winter cottage' close to his rich friends in the area of Santa Rosa and Altadena Drives to escape the snows of his New Jersey home. The gate house, a private residence now, features a stone tower with crenelations and fashioned capstones at the top and quoins highlighting each corner. Possibly Colonel Green felt the medieval English castle tower would protect his wood mansion.

"A flamboyant patent medicine magnate"[6], a *Los Angeles Daily News* articles relates, "'he made his multimillions...by developing real estate, like hotel Green' (in Pasadena on Green Street, of course)". The hotel was renamed 'Castle Green' around 1924 when it was converted from a hotel to private apartments. "He had a thing for castles," the Altadena Historical Society president Jane Brackman said. Possibly the Colonel considered himself a king among his fellow entrepreneurs.

WILKINS HOME
2500 ALTADENA DRIVE

At the bluffs of the Arroyo Seco sits a perfect river rock home, restored and maintained like few others. The attention to detail appeared in the smallest of places like the mailbox, the wrought iron light fixtures, and the

A 1985 photo of Colonel Green's Victorian mansion in Altadena showing the stone tower 'gate house' on the left.

blue-green trim. Excellent landscaping along with the gravel driveway, instead of asphalt, all illustrated the owners' sensitivity and sophistication. This was a home the owners were clearly proud of. The refined placement of the stones at the windows along with the thin mortar joints binding the stones are the signs of an expert mason, a mason who knows what to do, enjoys doing it, and is pleased of his efforts.

In 1929, Mr. Wilkins, a Pennsylvania stone mason, built his granite residence just before the stock market crash in October of that year, which may have destroyed his dreams but did not destroy his stone home. In a 2007 article, Ruth Ryan wrote, "the home was built in 1929 by Hugh Walker Wilkins...His wife, Mary, designed the house in Spanish Revival style, popular at that time."[7] Ryan may be forgiven for not knowing the difference between Spanish Revival and Medieval Castle styles, but 'Spanish' style resembles the stuccoed adobe missions, not river rock granite. Dr. Mary Quane, the Mary who designed the house, arrived in Southern California in 1922, thirteen

years later than Hugh. She may have been fleeing her two ex-husbands and a "passel of children she hardly knew."[8] Today, Mary would be proud of the thorough and sensitive restoration that took many years. Her home remains a rare and a perfect example of river rock buildings.

Though rare, there are more.

ARROYO SECO

Further south spanning the wide canyon, which provided the major source of the river boulders and river rocks for the local homes and churches, is the Arroyo Seco canyon. It begins in the San Gabriel Mountains and ends the Los Angeles River. At the base of the mountains Altadena was founded on the west side of this canyon and Pasadena on the eastern side. Connecting these two cities and spanning the gorge is the 150 foot tall Colorado Street bridge. Built in 1913 and improved by the WPA in 1933, it's a historical landmark popularly dubbed the 'Suicide Bridge' after the devastating stock market crash in 1929-30, a lifelong moniker which the City of Pasadena and its Chamber of Commerce wishes would go away. But this may not happen soon for over one hundred jumpers (two as recent as 2017) over one hundred years gave reason to keep that name. "...a number of spirits are said to wander the bridge itself as well as the arroyo below. Others have heard unexplained cries coming from the canyon. One report tells of a spectral man that is often seen wandering the bridge who wears (sic) wire rimmed glasses. Other people have claimed to see a woman in a long flowing robe standing atop one of the parapets, before vanishing as she throws herself off the side." And historical myths continue about the arroyo under the bridge, "...phantom forms have been seen walking the river bed, a number of unexplainable sounds are often heard, atmosphere is often described as 'thick.'"[9] Historically, a particular bent of people have been drawn to canyons with their private and quiet wooded surroundings. This wide, deep, and long canyon, the Arroyo Seco, has followed that heritage.

The south end of this long canyon of sycamores and oaks connects with the Los Angeles River. Quite a misnomer as 'river' presents a picture of treed gurgling cool waters bordered by smoothed stones. It is nothing of the sort, but now another dry Los Angeles concrete channel of graffiti and twenty-foot-high concrete walls, an outdoor prison for flood waters not seen for decades.

Pat Morrison with photographer Mark Lamonica wrote a book celebrating this river, *Rio LA* in 2001 which describes how, before its imprisonment between concrete walls, "The river ambled across a Los Angeles that was unimaginably wetter than it is today...wetlands and wooded lands thrived. Immense stands of sycamores and cottonwood."[10]

A FEW TECHNICAL THOUGHTS

Styles change, frequently with each generation. In the early twentieth century, people in the U.S. may have been in a visual hangover from the ostentatious Victorian filigree and the bombastic

Lower part of Arroyo Seco Canyon in 1932, before freeway and binding the river in a concrete channel.

Romanesque styles and sought simpler designs. The new bungalow style craftsman homes fulfilled this demand and became popular throughout the U.S. These simple homes were noted for their use of natural materials, stained shingle siding, and river rocks; their large overhangs shaded the home from the intense summer sun and covered verandas extending the full width of the house provided relaxing moments in the evening's cool air.

In the granite country of Southern California, river rock satisfied this naturalistic preference. But not all wanted the craftsman bungalow style. Some settlers dreamed of a medieval castle design of the type seen in the hills of Europe with crenelations lining a flat roof, or their wives may have dreamed of the Irish castles in her homeland with round turrets at one corner. Those with a more limited budget or practical nature, such as a farmer from Iowa or Missouri, chose a simple rectangular shape where function prevailed over aesthetics and a smaller budget limited its size.

The Keystone is an arrow shaped stone that is the key so that all the others in this arch stay in their places

The strength of these river rock buildings results from molten rocks, their slow cooling over millennia, and then binding the stones with Portland cement into a jigsaw puzzle arrangement. Each stone's placement interweaves with its neighbors, creating a cohesive fabric of strength and kaleidoscopic patterns of shape, color, and texture. You may notice the interlocking placement of the stones. This occurred intentionally, where one dips the other bulges, where one slopes off its neighbor is large and fills the empty space so the expense of mortar is minimized. A lot of mortar shows a novice at work; whereas, a master mason places one heavy stone at a time not ruminating over which stone to pick just knowing by sight and feel the right stone and the one place, frugal with his mortar. An artist.

Large gaps filled with mortar crack and spall, weakening the wall. Unlike the uniformity of bricks, these irregular sized and shaped stones do not allow the straight lines of brick construction. Irregularity bonded into uniform whole creates the enduring strength of a cobblestone wall. These technical understandings may help improve our appreciation of these stone homes and buildings and their makers.

Stone mason technology was not new. We all learned of it during our history classes-seeing pictures of France's towering cathedrals and the beautifully proportioned Greek temples. It was the Romans 2,400 years ago or so that invented the stone arch. "As one of the basic building blocks of architecture," Cristine Sciacca writes in her book Building the Medieval World, "the arch came into its own in the Roman period..."[11]

The next time you walk under a stone arch, stop, and look up. The keystone at the peak of the arch, that arrow shaped stone, was critical for the hundreds of pounds of stones to remain there for centuries. How does that work? you wonder. Those stones should fall down. But there they stay, even in earthquakes, under the stones' own laws of shape and gravity, cramming to get into a store's door when just opened for a black Friday sale. Their weight keeps each in place. Remove one and the arch collapses. This graceful placement of these stones weighing 20-40 pounds each requires skill.

Consider the mason as he worked. He may have spread out the stones on the ground to better select his next stone, the methodical way. Or he may have worked from a pile of stones selecting his next stone from its perimeter as an experienced mason could do, alert to color and pattern as the stone fit into the building. Artistic creations may begin with inspiration, but they all end with effort.

The masons, and the people who hired them, built with stones for many reasons. Many of those reasons will remain unknown, but can be surmised until first-hand accounts are discovered. I'm going to venture a few ideas and probabilities. A few, not many, individuals selected thick stone wall construction over the more prevalent thin wood wall construction because:

1. The stone walls reminded the home owners of 'back-home', as would occur with settlers from Ireland, England, German's Alpine villages, and the rocky countries of Italy or Greece. When making a new home in a new country, little brings more psychological comfort than a house that resembles your childhood homes.

2. Yes, the stones were there. That is a significant factor, but not the only one or the most important one.

Stone pillar in Pasadena, typical at the larger homes driveway entrances.

The stones had to be hauled from distant sources, usually a dry wash, as there would not be adequate volume or size of stones laying around on the building site.

3. Building with stone requires hard labor and persistence. Building with milled wood requires less labor and less time; wood, of course, was used to finish the interiors, the doors, windows, and roof construction. The floors of stone homes were either built of wood or concrete. Again, this depended on the owner's budget, abilities, and spouse's preferences.

4. The following reason to build with stone is by far the most important for practical people, and California's immigrants were very practical. Stone was fireproof, kept the winds and dust storms out, was vermin proof with no cracks for rattlesnakes or black widow spiders to crawl through into their home. Wood siding leaks dust and cold at every board joint. Wood needs paint. Wood rots. Stone doesn't do either. Stone requires no maintenance and is washed in every rain bringing out its subtle beauty of sparkling textures, patterns, and colors.

Books on home designs became readily available in the 1910s for the builder and owner to review.

5. Every stone homeowner we interviewed was proud of their unique home. Not many homeowners of stucco or wood boxes share that sentiment.

6. Stone homes are rare and may be valued more than their wood neighbor. Historically, architecture was used to reinforce a person's status. Stone castles in the European countries and England were seen as the highest status in the land. These European settlers grew up knowing this cultural belief.

7. They last. Stones just do not deteriorate; winds will not blow them away, like some wood buildings. Earthquakes have not toppled them like other buildings and bridges, as we will see in chapter 6.

These smooth, rounded, granite boulders are not mute for they dictate the rules of constructing the wall's shape, size, height, and thicknesses. Building with stones bound together with Portland cement left little margin for error. A mistake in the angle or length of a wall, the size of a window or height of a door that needed corrections meant tearing down that stone wall and rebuilding. Not an easy or welcome task.

Why didn't earlier settlers of Southern California build their structures of stone instead of adobe, as in the Hispanic missions, or of branches and foliage as the Chumash did with their conical homes? River rocks do not stack, like oranges or basketballs do not stack. A binder is needed to hold them together. And in the late nineteenth century that binder was found: Portland cement. Much stronger than either lime or gypsum-based cement, Portland cement brought from England provided the mortar technology necessary to construct stone homes.

Also from Europe came the masons who knew how to use stones, they arrived in New York and continued west, settling in the foothills of Southern California. This was familiar landscape for those from mountainous country. These two ingredients completed the vacuum so that the thousands of stones in dry river washes and in the ground of the alluvial fans became buildings.

How were these buildings constructed? We've reviewed the who and where, but not the technology of their construction. That's next.

END NOTES

1 Reed, H.A., *History of Pasadena*, Pasadena Historical Co., 1896.

2 Weaver, J., *Los Angeles the Erroneous Village*, Capra Press, 1980, p. 51-2.

3 The Living New Deal, www.thelivingnewdeal.org.

4 Dotson-Retna, L., "The Family Stories We Share and What They Mean to Our Children," in *The Washington Post*, November 22, 2019.

5 True, J.M., *Painted Rocks*, Arts and Crafts Press, 1965, p. 129.

6 "Llano del Rio," Wikipedia, retrieved August 18, 2018.

7 Ryan, R., "Chiseled Good Looks, " *Los Angeles Times*, October 28, 2007.

8 Rollins, P., "Magical Garden for a 1929 Spanish Colonial Revival," Arts and Crafts Homes and the Revival. artsandcraftshomes.com, June 5, 2012.

9 "Suicide Bridge, Colorado Street Bridge in Pasadena," Legends of America, www.legendsofamerica.com

10 Morrison, P. and M. Lamonica, *Rio L.A.*, Angel City Press, 2001.

11 Sciacca, C., *Building the Medieval World*, J. Paul Getty Museum, 2010, p. 76.

CHAPTER 5

River Rock Construction Techniques

"Why Stone? Well, why not? For building...you simply can't do better. Stone is weatherproof, rat proof, insect proof, and long lived...It signifies good taste."

Stonework by Charles McRaven, published by Storey Books, 1997

A mason or owner building a cobblestone home.

It's stone. Granite stone. Hard. Smoothed. Rounded ovals. And heavy. Just stacking river rocks is impossible, the stones roll off each other into a heap. You need to bring in Portland cement mortar; it's the glue that binds them together. When wet, Portland cement is the consistency of mud. Soft, squishy, gray, caustic. Temporary wood forms are also needed, and importantly, a mason's skill and the his tools. These will complete the items needed. After constructing their stone home in Idaho, Lewis and Sharon Watson "searched out and reviewed many books on stone masonry...The most striking fact of our research is that modern literature on old-fashioned stone masonry is essentially non-existent!"[1]

Though a little knowledge is dangerous, my forty-five years of architectural experience, many of those years working with century-old cobblestone or wood-framed buildings, and a little hands-on setting of river rocks for garden walls at home along with reviewing the scant literature available, prepared me to share these findings and understandings of the process to build a cobblestone structure.

AN OVERVIEW

After buying your dream lot, you may have discussed your future home's appearance and size with your spouse, including the number of rooms and their sizes, and their arrangement. Was a covered porch and single car garage desired? Should it look like an Irish castle or a stone arts and crafts design? You can imagine that many after dinner conversations to decide these issues, especially since this was most likely your first home you designed and built. It's your dream home.

THE SITE

Upon arriving in Southern California as a settler or visitor, you would have searched for the right location to buy. Selecting the site in the later 1800s depended on the source of water and on your wealth and health. In the early twentieth century, unimproved barren lots were for sale by land developers and real estate brokers. Land developers had bought large parcels from the Spanish and Mexican land grant holders to subdivide into smaller parcels, grade dirt roads, and provide water to each lot. Selecting a site then became easier.

Those lots were large, up to a half acre, 50 feet x 300 feet for the owner to raise animals and do modest farming. But first, you needed to erect a temporary home, either a tent on wood platform or build a simple wood cabin for you and your family to live in while building the stout stone home, which took about three to four months if you chose a simple design. Building was quite an effort, taking long days. You may have hired a mason or Mexican worker to help out with collecting the stones and mixing the cement.

BEFORE CONSTRUCTION

Simple rectangular or square shaped buildings with few openings and without offsets in the walls may not have required plans, but did require functional direction. Was the building for storage or a smokehouse or a home for your family?

Elevation sketches showing wall heights, opening sizes and locations helped determine the sizes of the stones needed and the thicknesses of the sloping walls. Special shaped stones with two flat faces at right angles were needed for the corners, thin, long stones for the head of the openings, and the arrowhead shaped stone for the keystone at the top of each arch. Finally, an approximation was calculated of the number of stones required to build the structure. Hauling up too many from the wash was wasted effort and you'd also need to determine what to do with the extra pile of stones.

Then mortar must be bought, tools assembled, water brought to the site, and wall lines laid out. Not a stone had yet been placed. To some would-be masons this would be arduous and had to be learned (think of driving to the store but not knowing how to drive). While to a master mason this was simple task, like a race car driver at Indianapolis."One part skilled craftsman," an article in OldStoneHouses.com relates, "and one part artisan, the stone mason could envision the design, place the stones, and then set them in a tightly woven, incredibly intricate pattern that required little mortar to stay put."[2]

First though you'd need to prepare the building site. That could be easy or intense. The site was to be selected on your lot, either towards the front, near the street, or set back. Was there a view or prevailing breezes towards which to orient windows and porches? On sloping land the site must be leveled unless you were building a basement and the floor was wood- framed. In that situation, a deep hole would have been dug and the stones found in the soil cleaned and set aside to build the basement walls. At some time while you were digging the hole, it was very possible that you thought, this basement doesn't have to be the size of the house floor, just big enough for storage of our jarred food, to hide the wine (during Prohibition years), and to store yard tools. The smaller size was chosen, the dirt floor was leveled, the formerly dirt stairs were paved with stones or concrete, and the stone walls were built.

If a basement seemed either too much effort or not needed, the site needed clearing of brush and boulders, leveled for a packed dirt or poured concrete floor, which included building low boulder retaining walls. But, if you had the resources to buy extra wood to build joists and posts for a subfloor and laid clear pine or oak planks for smooth finish floors, then the land could slope and remain ungraded. But if you were building during the 1920s after WWI when plumbing and electrical work became standard, the drain and supply lines along with any electrical wiring under the floor must first be installed along with connecting these plumbing lines to wells, cisterns, septic tanks (for all of which large holes had to be dug) or connect to them to the water pipes and electrical lines located at the dirt street. There was a lot to learn and you would have checked in with neighbors and building supply yard for advice and instructions.

ASSEMBLING THE MATERIALS AND TOOLS

Gathering the needed tools and materials occurred before and sometimes while laying out the building on the site. The tools included: twine, wood stakes

A display of mason's tools at Bolton Hall in La Crescenta

(sometimes made on-site), lumber for scaffolding and for the forms for window and door openings, ladder, nails, Portland cement, some river rocks and boulders, and, most necessary, water. A few weeks were needed for the windows to be built at the lumber yard, so you would order them early, but the window forms had to be built first to determine the sizes.

The mason's tools included not only shovels and spades for clearing the site of brush then leveling the ground, but cutting tools like a scythe and saw, also hammers, trowels (usually two types, a larger one for troweling the mortar and a small trowel for pointing or pushing mortar into small joints), rubber or wood mallet for hitting the larger stones into place, and dense rags made from burlap potato sacks. To align the wall corners, a builders square that is a right 90-degree angle wood triangle using 3:4:5 dimensions on the sides, was built. The vertical alignment of the inside face of the wall was easy, just tie a stone to twine and let it hang.

PORTLAND CEMENT MORTAR

Mortar's unique properties can't be seen. They are chemical. Those chemical reactions occur as the mortar dries as it changes from a muddy substance to hard glue binding the stones together. Mortar filled the voids as well as bound the stones, resulting in a solid monolithic wall. Without going into a chemistry lesson, just know that a special kind of mortar, invented in the late 1800s, not only resists erosion but binds stronger. It is Portland cement.

Wind and rain erode a more common lime and sand mortar. In time, it deteriorates and crumbles to where you can rub out the mortar joint with your finger, turning it to sand. This type of mortar is found in old brick buildings, such as commercial and institutional buildings, where costs are watched closely, as this mortar is inexpensive. Over the decades the ingredients in mortar turned into a science. But a century ago you mixed the Portland cement in with sand and lime and then added water to mix the mortar to get the desired consistency: dry mortar to patch holes or wetter, like thick mud, to set stones.

THE PROFESSIONAL MASON

Laconic. Even tempered. A man of few words, so few that in our years of research only one mason wrote of his methods. A stone mason at work thinks things through, plans ahead, is practical. For making an error in a stone wall is difficult to fix, especially if the mortar has dried. There are few highs, few lows in his speech, and he's spare with the use of adjectives and less of adverbs. Mostly nouns, some verbs. Though his stone work shows artistic talent you will not hear aesthetic pronouncements from him, just the advantages and disadvantages of different techniques or approaches. His artistic talent emerges when in the zone of selecting and placing the stones in the mortar.

Master masons constructed with durableness in mind, and used simplified ways that spared their labor and decreased their efforts. They didn't skimp, they thought. He might have said to you after you asked how he works, "This is a cobblestone. You don't have to be a genius to fit them together." Humble yet proficient, experienced.

An example of a stone arch above a window. Note the large amount of mortar between the stones and the gaps between the wood ceiling boards and the stones.

CONSTRUCTION

A professional stone mason knew from experience that the completed river rock structure acted as one solid piece, like a car or airplane: one structure assembled from many parts. Deep foundations were not needed as the wide base of the wall along with the virgin soil, compacted for many centuries, provided excellent support.

The building's perimeter was staked out on the cleared and leveled site and taut twine was strung to show the wall alignment. Door and window openings were located. Before constructing the wall, the stones were gathered and placed on the inside of the building. The mason would have sorted special stones from the wash and set them aside to be used in special locations, such as the top of an arch, or using larger boulders for the base of the walls, and for the corners of the wall that required stones with naturally occurring two flat faces at right angles. Each must be a particular shape. And during the summer months the mason would feel the heat in the stones and during the winter embrace the cool stones for these bind with the mortar slowly and deeply. In the summer, warm to hot stones dry the mortar more quickly resulting in weak bonding. Each session and daily weather affected the mortar mix, drying time, and number of stones placed. Then when all was readied for the cobblestones to be set in the correct locations, the right amount of mortar was mixed, enough for up to an hour's work, too much and the mortar dried making it stiff, useless, and wasted. These were frugal years.

The huge stone above the hearth, its markings like a Japanese painting of clouds, its dark charcoal tone like an ink brush painting. This boulder was specially selected to be placed in a prominent position above the fireplace.

Beginning at a corner, the mason shoveled a thin layer of mortar in the shallow trench to make a mortar bed and set the large base stones in, grouting the sides and voids. As the walls grew in height, scaffolds, ramps, and pulleys to place the heavier stones would be built when needed. Wooden ramps allowed the workers to push a heavy boulder five feet or so above the floor. Using pulleys exposed the workers to the danger of the boulder falling out of the sling or platform used to lift it. Both had their dangers and probably a few damaged fingers occurred.

The mortar was mixed to the consistency of cake batter, so when the stone was placed the mortar squished around the stone's irregular shape, and then the mason tapped it into place with the end of his trowel. A final flick of his wrist, a quick and natural sideways movement, and the trowel scraped away the excess mortar. A firm joint was made clean.

While building the walls, the mortar joints between the stones needed finishing. A right way to finish them would be to wait one to two hours after placing the mortared stones, the time depending on the weather, its humidity and temperature, sun exposure, and —all affecting drying time for the Portland cement to set (cure); then with a gloved hand or moist burlap gunny sackcloth, you rubbed the joint to even and smooth the mortar. Or smoothed it with a paint brush. Not done yet, there was one more step. Using a different wetted cloth the joints were wiped to a toothpaste-even finish and the faces of the cobblestones were wiped clean of excess mortar. [3]

WOOD FORMS

Wood forms were used to build stone walls for they provided vertical alignment to set the river rocks on the inside face and provided temporary support at openings. These supports were needed until the mortar dried and the stones were held in place, especially for arches. This was called 'forming'. Usually 2x6 or 2x8 wood boards eight to ten feet long are placed horizontally on the inside face of the wall, stacked two high, and braced by supports staked into the earth. These backer boards could be horizontally or vertically framed and moved down the wall as you progressed. Other methods such as 'slip forming', required forms on the front and back sides of the wall, and a third method was to not use forms.

But arches at windows or porches required a different type of wood form. The mason asked a carpenter to make one form that could be used repeatedly. They would be built in the shape of the opening, the form's curved arch at the top was braced with 2x6's to prevent sagging and the straight vertical sides extended to the window base. Thus, the mason knew their size, and to make life easier, for masons were always looking for ways to simplify their jobs, all or most of the windows were the same size and shape, one form moved down the wall to be reused for many openings. Many windows were the same

The wood forms used in construction leave their impressions.

APRIL 14, 1922. SOUTHWEST BUILDER AND CONTRACTOR 27

CURRENT PRICES OF BUILDING MATERIALS

LUMBER

lumber at Los Angeles yards, which became effective Oct. 15, 1921, have not since been changed, and are the only officially published prices:

1-in. Common Douglas Fir, rough (Oregon Pine)	6 to 14 ft.	16 to 24 ft.
1x3 Per M	$35.00	$37.00
1x4 Per M	35.00	37.00
1x6 Per M	37.00	39.00
1x8 Per M	37.00	39.00
1x10 Per M	37.00	39.00
1x12 Per M	39.00	41.00

For selected common add $5.00 per M.

2-in. Common Douglas Fir, Rough	10', 12', 14'	8, 9, 16' to 24'	26' to 32'
2x3 and 2x4	$35.00	$38.00	$40.00
2x6	35.00	38.00	40.00
2x8	35.00	38.00	40.00
2x10	35.00	38.00	40.00
2x12	36.00	39.00	41.00
2x14	41.00	44.00	46.00
2x16	42.00	45.00	47.00

For selected common, add $5.00 per M.

3-in. Common Douglas Fir, Rough	14' and under	16' to 24'	26' to 32'
3x4	$39.00	$42.00	$44.00
3x12	41.00	44.00	46.00
4x4 and 4x6	41.00	44.00	46.00
6x6, 6x8 and 8x8	41.00	44.00	46.00
8x10, 10x12	41.00	44.00	46.00
10x10, 12x12	41.00	44.00	46.00
14x14	44.00	47.00	49.00
16x16	44.00	47.00	49.00

1-in. No. 2 Clear and Better Douglas Fir, Rough	6 to 12 ft.	14 to 20 ft.
1x3	$70.00	$72.00
1x4	70.00	72.00
1x5	83.00	85.00
1x6	75.00	77.00
1x8	80.00	82.00
1x10	80.00	82.00
1x12	85.00	87.00

For Straight No. 1 Clear, add $5.00 per M.
For wider stock add to 12" price $5.00 for each additional 2"; for 1¼" and 1½", add $5.00 to price of 1"; for Vert. Grain, add $7.00 per M.

2-in. and Thicker, No. 2 Clear and Better, Douglas Fir, Rough	14' and under	16' to 24 ft.	26' to 32 ft.
2x3 and 2x4	$60.00	$63.00	$70.00
2x6, 2x8, 2x10	65.00	68.00	75.00
2x12, 3x4, 4x4	70.00	72.00	80.00
3x12, 4x6	75.00	78.00	85.00

For straight No. 1 Clear (on above sizes only) add $5.00 per M.; for Vert. Grain, add $7.00 per M.
For straight No. 1 Clear (on above sizes only) add $5.00 per M.

No. 2 Clear and Better	14' and under	16 to 24 ft.	26 to 32 ft.
6x6, 6x8, 8x8	$80.00	$83.00	$90.00
10x10, 12x12	85.00	88.00	95.00

Stepping S1S and Nosed.	6 to 12 ft.	14 to 20 ft.
1¼x10, No 2 Clear and Better	$ 95.00	$ 97.00
1¼x12, No. 2 Clear and Better	100.00	102.00
1¼x14, No. 2 Clear and Better	105.00	107.00
1½x10, No. 2 Clear and Better	95.00	97.00
1½x12, No. 2 Clear and Better	100.00	102.00
1½x14, No. 2 Clear and Better	105.00	107.00

For Straight No. 1 Clear, add $5.00 per M.

Douglas Fir Ceiling	6 to 12 ft.	14 to 20 ft.
1x4	$60.00	$62.00
1x6	65.00	67.00
5/8x4	58.00	60.00
5/8x6	63.00	65.00
½x4	55.00	57.00
½x6	60.00	62.00
3/8x4	45.00	47.00
3/8x6	50.00	52.00

For straight No. 1 Ceiling add $5.00 per M on 1" and $2.00 per M on thinner.

Douglas Fir Flooring	No. 1 Clear 6to12 ft.	No. 1 Clear 14to20 ft.	N. 2 Clear 6&12 ft.	N. 2 Clear 14to20 ft.	N. 3 Clear 6to12 ft.	N. 3 Clear 14&16 ft.
Vr Gr 1x4	$85.00	87.00	80.00	82.00	60.00	62.00
Vr Gr 1x6	90.00	92.00	85.00	87.00	.....	.....
Vr Gr 1x3	95.00	97.00	90.00	92.00	.....	.....
Vr Gr 1¼x4	95.00	97.00	90.00	92.00	.....	.....

	No. 2 Clear and Better 6 to 12	No. 2 Clear and Better 14 to 20	No. 3 Clear 6 to 12	No. 3 Clear 14 & 16
Flat Grain 1x4	$60.00	$62.00	$55.00	$57.00
Flat Grain 1x6	65.00	67.00	60.00	62.00

Worked Interior Finish, including Moulded Base, Casing, Apron, etc., or S4S and Sanded Stock.
No. 2 clear and better, (not over 12" in width), 6 to 12 ft., $100; 14 to 20 ft., $102.00, for 1x¾" and over, add $5.00; for straight No. 1 Clear, add $7.50 per M; for Vertical Grain, add $7.00 per M; for Wider Stock, add to 12" price $5.00 for each additional 2".

Redwood Ruf, Common	14' & under	16 to 20 ft.
1x3	$47.00	$49.00
1x4	44.00	46.00
1x6 and 1x8	47.00	49.00
1x10	50.00	52.00
1x12	55.00	57.00
2x3	50.00	52.00
2x4	47.00	49.00
2x6	47.00	49.00
2x8	47.00	49.00
2x10	47.00	49.00
2x12	55.00	57.00
3x4	49.00	51.00
3x12	55.00	57.00
4x4 and 4x6	55.00	57.00
4x12	55.00	57.00

Redwood Rough Clear	12' & under	14 to 20 ft.
1x3 and 1x4	$78.00	$82.00
1x5	83.00	85.00
1x6	86.00	90.00
1x8	81.00	83.00
1x10	83.00	86.00
1x12	86.00	88.00
1x14	91.00	93.00
2x3	76.00	79.00
2x4	79.00	84.00
2x6	83.00	88.00
2x8	79.00	83.00
2x10	81.00	84.00
2x12	87.00	90.00
2x14	90.00	92.00
3x4	81.00	86.00
3x12	91.00	93.00
4x4	81.00	86.00
4x6	88.00	90.00
4x12	92.00	94.00
6x6	93.00	95.00
8x8	94.00	96.00
10x10	105.00	107.00
12x12	116.00	120.00

Over 14" add $6.00 per M for each additional 2"; for 1¼" and 1½" CLEAR, add $10.00 per M to price of 1".

Redwood Ceiling		
3/8x4 and 6 random lengths 7' and up	52.50	59.50
½x4 and 6 random lengths 7' and up	60.00	64.00
1x4 random lengths 7' and up	84.00	81.00
1x6 random lengths 7' and up	91.00	88.00

1x4 and 1x6 6' and under $5.00 per M less.

Redwood Battens, 3/8x3 ruf	47.50	45.50

Redwood Bevel Siding	No. 1.	No. 2.
½x4 from 1-in. random lengths	$47.50	$45.50
½x6 and 8 from 1-in. rand. lengths	49.50	47.50
5/8x4 from 1¼-in. random lengths	65.00	63.00
5/8x6 and 8-ft. from 1¼-in. ran leng.	66.00	64.00
Redwood Rustic—random lengths	90.00	87.00

Redwood Interior Fin., 12 and under $100.00
Over 12" add $10.00 per M for each additional 2".

Brackets, Rafter Ends and Verge Boards, sawn to detail $100.00

Shingles	No. 1	No. 2
Redwood, per bdl.	$1.20	$ .85
Red Cedar, Star A Star, per bdl.	1.50	

Shakes—6x36 sawn	$40.00	$38.00
6x36 split	40.00	......

Posts—4x5x7 Split Redwood $0.40
Short and Narrow in proportion
Posts—4x5x7 Split Redwood $0.45
Cartage not less than $2.00 per M.

Mill Charges.
Surfacing 1 side, range from $3.00 to $5.00 per M; 2 sides, $3.50 to $6.00 per M; 3 sides or 4 sides, $6.00 to $8.00 per M. B M; T & G or shiplap, 1 and 2-in., $5.00; over 2-in., $7.00. Sanding, $7.00 per M; Ripping, 2-in. and under, per M. lin. ft., $2.00; Resawing, each cut, per M. B M. $6.00.

BRICK AND TILE

	Per M
Common Red Brick, delivered	$ 15.00
Selected common red, delivered	17.00
Blue Brick, delivered	17.00
Pressed brick, red, f o b factory	40.00
Pressed brick, buff, f o b factory	50.00
Pressed brick, cream, f o b factory	50.00
Enam Brick, headers & stretchers, f o b fac	95.00
Enamel Brick, quoins, f o b factory	105.00
Hollow Tile, 2-in. 12x12 in., f o b factory	95.00
Hollow Tile, 3-in. 12x12 in., f o b factory	80.00
Hollow Tile, 4-in., 12x12 in., f o b factory	95.00
Hollow Tile, 6-in., 12x12 in., f o b factory	123.00
Hollow Tile, 8-in., 12x12 in., f o b factory	176.00
Hollow Tile, 10-in., 12x12., f o b factory	220.00
Heath Unit Tile, f o b factory	95.00
Universal Unit Tile	100.00
Roof Tile, red Spanish, per sq. fob fac. small	15.00
Roof Tile, red Spanish, per sq. fob fac, large	15.00
Roof Tile, red Granada, per sq. fob fac.	18.00
Roof Tile, Granada, vari-colored	21.00
Roof Tile, red Mission, per sq. f o b fac.	21.00

SAND, GRAVEL, CRUSHED ROCK

Price per ton at quarry, freight to be paid by purchaser (freight from quarry to bunkers, Los Angeles, 95c ton):

Sand (clean)	$ [illegible]
Gravel (screened rock ¼-in. to 3-in.)	[illegible]
¼-in. Crushed Rock	1.00
¾-in.-1-in. Crushed Rock	[illegible]
1½-in. Crushed Rock	.90
[illegible] Crushed Rock	[illegible]
Rock Dust	.50
L. A. pit sand or gravel, at pit, per yd.	1.10

LIME, CEMENT, PLASTER—Warehouse Prices

Cement, gross, f. o. b. Los Angeles	Per bbl.	$3.90

Rebate for returned sacks, 15c each.

Lime	Per bbl.	[illegible]
Plaster (sack refund 12c each)	Per ton	21.00
Plaster	Per sack	1.05
Wood Lath	Per M	10.00
Metal Lath	No. 27 gauge	.25
Metal Lath	No. 24 gauge	.29
Buttonlath, interior	Per yd.	.34
Buttonlath, exterior	Per yd.	.40
Buttonlath, double	Per yd.	.48
Lock-Lath, interior or exterior	Per yd.	.35
Corner Beads	Per lin. ft.	.05

CEMENT WORK

Base prices for cement work, for average conditions, as submitted by the Cement Contractors' Association:

Flat cement work, 4" thick (3½" base, ½" top) @ 20c per sq. ft.
City sidewalks (cement work only) @ 30c per sq. ft.
Straight Caps @ 75c per. lin. ft.
Straight Steps @ 75c per lin. ft.
Concrete foundations:
Bungalows @ 80c per lin. ft.
Two-story residence @ 70c per cu. ft.
Industrial and Heavy foundations @ 60c per cu. ft.

STRUCTURAL STEEL

Following are the base prices per 100 lbs. at Pittsburg and the prices delivered on the job in Los Angeles: Pittsburg Los Angeles

The 1922 Costs of Building Materials

size, and in a colonnade forming a front porch, the columns and arches were the same width, height, and shape.

THE WALL

Full river rock walls require width. Depending on different conditions, like its height, the size of the stones, and the mason's skill, stone walls may be 24 to 36 inches wide at the base and taper to 9 to 12 inches at the top. Self-supporting stone walls were identified by their tilt or slant inwards on the outside of the wall showing this taper. And the river rock sizes changed as the mason progressed with smaller lighter ones used in the upper portion of the wall and the larger much heavier boulders at the base supporting the huge weight of the wall. Thirty-six inch thick walls were built with a handsome stone on the outside, and a flat faced stone for the inside with the core between these two usually filled with rubble and broken pieces of stone.

Now veneered stone walls, the type I call thick wallpaper, do not taper, they maintain the same thickness full height. Small stones and flat-faced stones indicate veneered stone construction where stones would be mortared directly onto the wood-sheathed walls that support the stones. Stuck on stone, I call it. For example, try stacking marbles and they fall-down; they need glue and a supporting back to stack vertically. Many buildings in California have been built with this style.

This photo of a desert rock building reveals how stones were stuck on the wood wall like thick wallpaper, phony like a 1920s stage set lacking three dimensional depth and authenticity. Some of the buildings in this book are the 'veneered' type, though we wished to focus on the rarer type of stone buildings with thick self-supporting stone walls, these veneered buildings snuck in due to other special features.

Most of the buildings presented in this book are self supporting stone walls 12-to-36-inches thick. Common to all river rock structures was both the physical stamina used to heft the 20-40-pound stones up to six times before setting the

A profile of veneered stone construction on wood framing. First, nail on the wood, then the wire mesh, then trowel on a thin coat of mortar, then while wet push the flat stones into the mortar and cement up the joints. This unfinished miner's house used stone veneer to provide resilient protection from the Mojave Desert's heat and winds.

stone in its final resting place, and a mason's experienced eye for selecting the right stone shape and color for the right place. It's like assembling a jigsaw puzzle. Instead of small cardboard pieces, the mason assembled a thick stone wall requiring two to four stones per square foot for each side of the wall. You and your mason worked your way down the wall from the corners and around to the other sides, each wall rising at the same pace, about three feet of height per a twenty-foot stretch in good day's work.

Upon finishing your modest home of six hundred square feet, a common size in the Southern California's rural communities in the early 1900s, you and your helpers would have lifted and set in place around 200,000 to 300,000 pounds of granite river rock, plus the Portland Cement mortar binding them and filling the voids. That is like lifting one blue whale or four humpback whales.

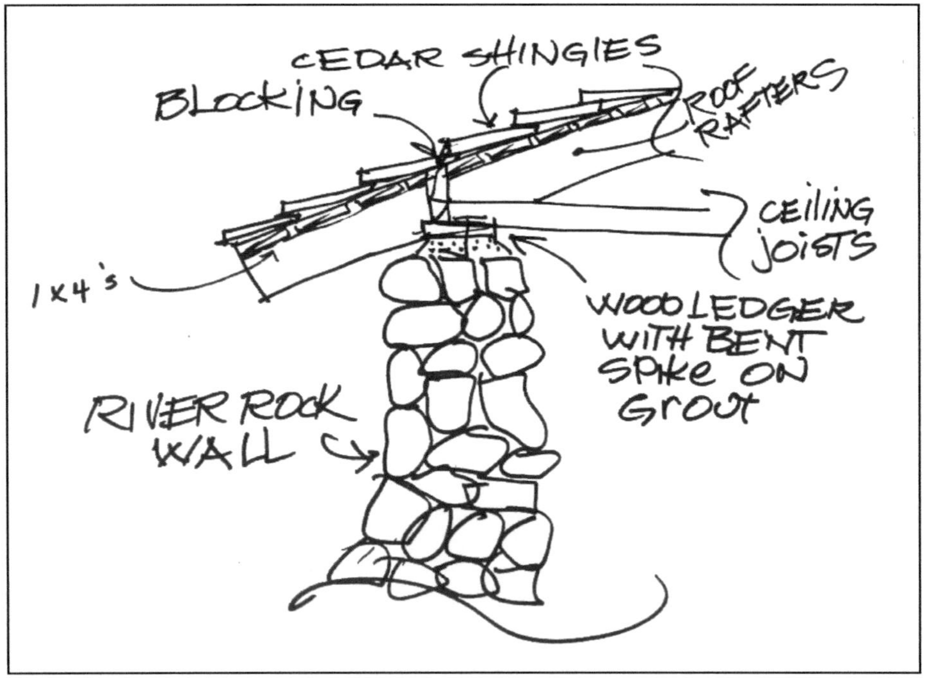

You have learned that using river rocks from a dry river wash saves time and labor as they would be clean, washed for centuries in the seasonal river. When dug up from alluvial soil, however, rocks must be washed by a laborer in water, a precious resource, to clean them so the mortar would stick to the stone and not the dirt. Mortar will not bind to dirty stones, sort of like painting beach sand. Your stones were placed in the center of the building. This makes your work easier and quicker. Lewis Watson, who wrote of his experience in building his stone home tells us that "It's wise to unload your rocks inside the intended house. That way you can build all four walls from the greatest selection of stones and generally carry each stone the shortest distance."[4]

ROOF CONSTRUCTION

Reusing the wood planks from the scaffolding and

873 Palmello, Claremont

Top: ***1003 Soldano Avenue, Azusa***
Bottom: ***902 Olive Street, Burbank***

forms, the carpenter built the roof framing. Simple and safe trusses would span thirty to forty feet, but the more common span was not as great. He chose the type of truss based on the span and his experience.

The critical issue was connecting the wood truss to the rock wall. To be able to fasten them required a transition, and it was a 'sill plate' Using a 2x6 plank about eight feet long, spikes were nailed all the way through the board and then bent. Setting this top sill into a bed of mortar on top of the wall, with those protruding bent spikes pushed into the mortar, securely fastened and leveled it to the stone wall. After the mortar set the roof framing was nailed to the sill. Redwood planks were always used for this sill plate as it did not rot. Less expensive Douglas Fir or Cedar as well as redwood was used for the roof framing.

And with all the trusses in place, 1x4 spaced slats were nailed perpendicular to the trusses to provide the supporting framing members for nailing the wooden roof shingles. Cedar was the preferred wood for shingles until the invention of asphalt shingles in the 1920s. What's fun about the cedar roof shingles was that during the dry summer months light filtered through spaces between the shingles; you could see small slits of the sky between the wooden shingles. But give them a little moisture and the wood swelled into a watertight roof.

THE FLOOR AND INTERIOR FINISHED WALLS

In most buildings a wood floor was installed two feet or so above the ground. Again wood from the scaffolding and the forms was recycled, but that would not be enough wood so more would need to bought from local lumber yard. Using modest dimensional lumber, 2x6s for floor joists supported by 4x6 beams about six feet apart with 4x4 wood posts also spaced about six feet that rest on wood planks or flat boulders, this rough framing supported the finish floor of either sanded pine or oak, or if money was available, then stained maple or walnut was installed. Both were expensive hardwoods.

As with the roof, to connect this floor framing to the rock walls, a spiked redwood sill plate was embedded into a bed of grout on a narrow stone bench at the base of the wall.

When concrete became readily available in the

mid-1920s, it became popular to pour a concrete floor, which required less labor, though this method was not faster than the wood-framed floor as the concrete must set, dry, and harden for at least two weeks before setting furniture or pianos on it. A concrete floor also had downsides: concrete doesn't move, it is rigid, and cracks appear, is colder in the winter which required more heat from your fireplace and standing for a while on its hard surface tires your legs. Plus, dropped china or glass broke every time. None of these conditions existed with wood flooring.

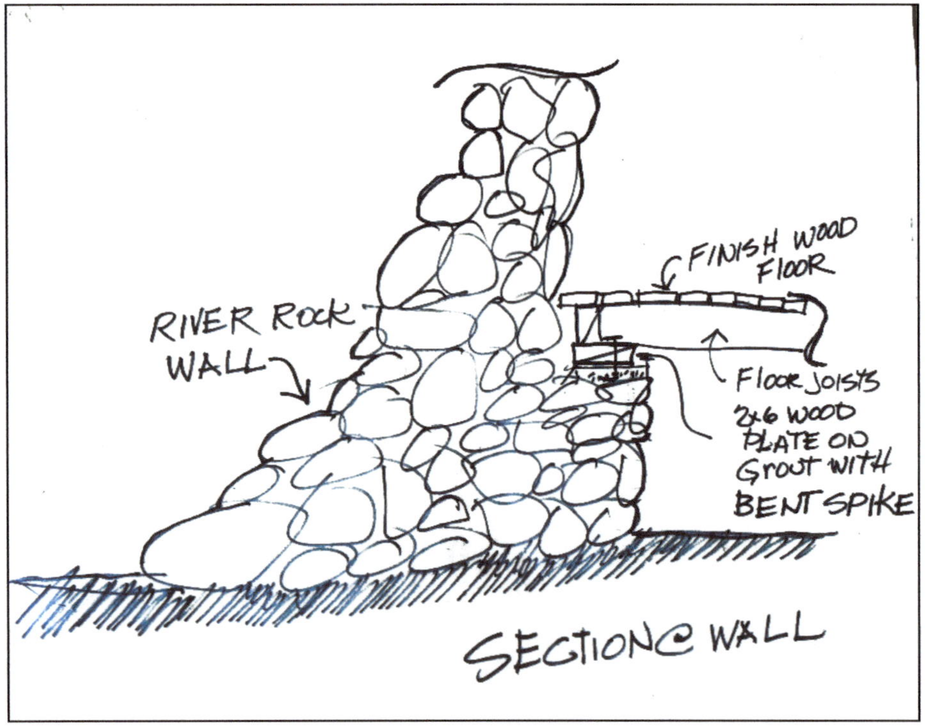

The interior was finished in a different ways, depending on the wife's influence and the family's budget: plaster on wood lathe supported by vertical 2x4 studs, named furring or a furred wall meaning not structurally supporting any weight like the roof, resulting in a smooth finish; or plaster applied directly to the stone for a lumpy finish; or wood paneling nailed to wood studs spaced from the rock wall an inch or so; or the stone was just left exposed (the preferred style for bachelors, garages, and basements).

QUALITIES OF RIVER ROCK BUILDINGS

Consider these aspects: river rock buildings use much less manufactured material which resulted in using much less energy (both electrical and fossil fuels). They require less air conditioning and heating thus another reduction in electrical and fossil fuel use. And they require little maintenance where only the wood portions such as door, windows, trim, and roofing need repair. Cobblestone construction has survived fire, floods, seismic and hurricane forces, as proven by their decades of exposure to them. They are free of drafts, impervious to dust, and mute city noises. Neither a 'concrete icebox' in chilly months, nor a 'stone oven' in the summer, river rock homes' thick walls maintain even temperatures as the walls heat or cool very slowly due to the thermal lag of the granite.

Now that we've built a house, you are ready to move in. And we move on to visit cobblestone homes in a number of neighborhoods.

END NOTES

1 Watson, L. and S., *How to Build a Low Cost House of Stone*, Stonehouse Publishing, 1974, p. 37.
2 Old Stone Homes, www.oldstonehomes.com, viewed August 8, 2018.
3 Haduck, M., "Stone Masonry," YouTube., viewed October 28, 2018.
4 Old Stone Homes, www.oldstonehomes.com, viewed August 8, 2018.

CHAPTER 6

Highland Park, The Arroyo Seco Canyon

"If we all had time to see the beautiful, it was ever ready before us, we should all learn to love it. Once the soul awakens to the spell, we cannot come away without a pang."

Charles Sumner Greene

In Los Angeles at the turn of the twentieth century, John Weaver wrote, that tourists would have found a "conservative, provincial, Protestant (and Methodist) open shop (non-union)" a small city that "makes do with meatloaf and buttermilk secure in the knowledge that temperance and industry would be rewarded with prosperity in this world, salvation in the next."[1] Expressed more than a hundred years ago, these thoughts continue to influence us. In 1909 a Municipal Arts Commission reported that, "The portion of the city which might be most beautiful and picturesque—all that northern section (today named Chavez ravine, Highland Park, and Echo Park) with its chain of rolling hills—is being truly ruined through the mistaken greed and ignorance of real estate speculators."[2]

As an example "To illustrate the dizzying rise of downtown (L.A.) real estate value," the *Los Angeles Times* in 1925 traced the history of "some land between 6th and 12th Streets and Main to Figueroa Streets. In 1864 Orzo Childs had bought the area (totaling about 30 city blocks in downtown L.A.) for $1 and dug a water ditch to irrigate his orchards...By 1924 their estimated value was $1,650,000"[3]

That land had appreciated 164,900 percent in half a century. Such criticisms of the conditions the growth of L.A. created did not end there. Weaver wrote, "Angelenos in 1909 were talking about downtown traffic jams (not much has changed), the need for a central library, a new ordinance shunting smokers to the back of the street cars (a lot has changed there)...the tenfold increase in in the cost of collecting garbage (and the increases have not stopped), a Packard dealer's proposal to connect Los Angeles and San Francisco with an automobile highway, and the soaring salaries of city employees, some of whom were knocking down $300 a month."[4] (At a 2017 level, this

Downtown Los Angeles in the early twentieth century looking west from Bunker Hill.

would be $8,100/month.)

Just north of central Los Angeles an intellectually and artistically oriented community settled in that beautiful northern section, and Highland Park was born. Occidental College, founded in 1887 in downtown Los Angeles, relocated to Highland Park in 1913, and needed housing for its professors. Sycamore Terrace filled some of that need. This hidden narrow street, known only to its residents and the mailman, is located on a hillside thirty to forty feet above the adjacent Figueroa Street, a major thoroughfare to downtown Los Angeles. Sycamore Terrace Street, lined by groves of Sycamore trees, became known as 'professors row'.

4939 SYCAMORE TERRACE

Made with the granite cobblestones from the Arroyo Seco, a few blocks to the east, this cobblestone home with its 22-inch thick walls contrasted with its traditional wood-sided neighbors. It was designed and built in 1892 for Occidental College Vice President Elizabeth Gordon Hunter from England. A century later it became designated by Los Angeles as Historic-Cultural Landmark #373 and remains in perfect condition, as if it were built yesterday. One feels the professorial stateliness and the intellectual discussions that most certainly occurred in this refined home.

An original scenic frieze above the dining room window recalls the bucolic years of California before miles of asphalted roads, hundreds of dull strip malls, and millions of look-alike houses defaced this landscape.

"This stone home," Neal Hammond said, "seems so much bigger than us, putting things into perspective. We feel the history, feel the people who lived here; we respect the house a lot." Sensitive musicians, Neal and Lucille Hammond are both concert violinists and the home's current caretaker-owners. They were delighted to find that the thick stone walls provided an ideal quiet environment for recording their music.

"It was a wonderful surprise for us, as we record and practice in the room closest to Figueroa, which is quite a noisy street." Neal continues, "Well before we repair anything we ask the home what it would accept. May sound a bit odd, but we really respect this house, its age, and its stones. And I sense when something is the right or wrong way to go." They praised the century-

This attractive house was built as a residence for Occidental College instructors by Elizabeth Young Gordon, the wife of John A. Gordon. The street, then named Fredonia and now Sycamore Terrace, became known as "Professors Row." The house is listed as Historic Cultural Monument No. 373.

Here are some shots of the original architectural details inside the house on Sycamore.

old home with its nine-foot-tall ceilings, the clear heart wood trim, and the oak flooring. "We love living here, it's beautiful. When the weather is hot it is way cooler and vice versa. We respect the natural beauty of the home." Neal compares living in this stone home to playing centuries-old violins. "Each has their character, resonance, and tones."

In the late 1800s, the Arroyo Seco offered inexpensive land and three other artistic individuals came to this 'more remote' canyon filled with sycamore trees to build their dreams with the canyon's stones.

THE HINER HOME

"...for Mama and I built this all by ourselves. We drove over into the Arroyo with an old car and brought all the stones for it, and then helped build it, too."[5]

Half a block south along professors row at 4757 North Figueroa Street is "The House that Love (and Music) Built" in 1922. Troy and Heather Evan bought The Hiner Home in 1966 then began immediate restoration. They cherish their English Tudor rock home restoring and repairing it like an antique car or a rare painting.

When the original owner, Dr. Hiner and his wife Anna, had arrived in Los Angeles around 1919, they were already in the middle of two long and distinguished careers. Doc Hiner, born in Kansas in 1871 or 1872, depending on the source, was not a young man when he and Anna began collecting the arroyo stones, for he turned fifty when he obtained the permit to build their home. Doc Hiner "performed many years as a noted Kansas City band leader," Lee Shipley recalled, Dr. Hiner also "led his Third Regiment Band to the Spanish American War." Remembering his childhood years, Shipley, a native of

Beginning construction of the Hiner home. Note the pile of stones at the corner of the home.

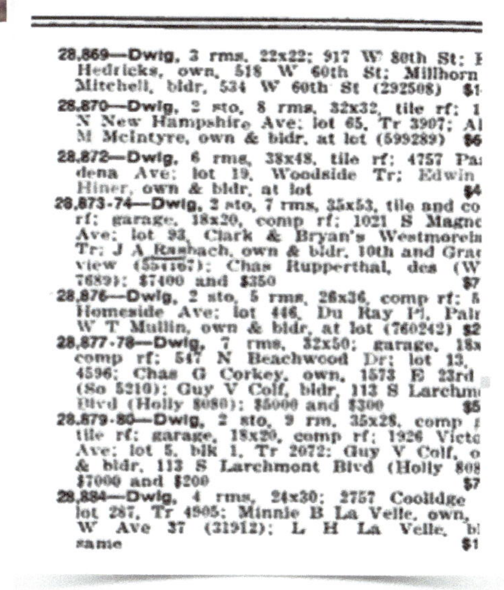

28,869—Dwlg. 3 rms, 22x22; 917 W 80th St; F Hedricks, own, 518 W 60th St; Millhorn Mitchell, bldr, 534 W 60th St (292508) $1

28,870—Dwlg. 2 sto, 8 rms, 32x32, tile rf; 1 N New Hampshire Ave; lot 65, Tr 3907; Al M McIntyre, own & bldr, at lot (599289) $6

28,872—Dwlg. 6 rms, 38x48, tile rf; 4757 Pasadena Ave; lot 19, Woodside Tr; Edwin Hiner, own & bldr, at lot $4

28,873-74—Dwlg. 2 sto, 7 rms, 35x53, tile and co rf; garage, 18x20, comp rf; 1021 S Magno Ave; lot 93, Clark & Bryan's Westmorela Tr; J A Rasbach, own & bldr, 10th and Gran view (554167); Chas Rupperthal, des (W 7689); $7400 and $350 $7

28,876—Dwlg. 2 sto, 5 rms, 26x36, comp rf; 5 Homeside Ave; lot 446, Du Ray Pl, Palr W T Mullin, own & bldr, at lot (760242) $2

28,877-78—Dwlg. 7 rms, 32x50; garage, 18x comp rf; 547 N Beachwood Dr; lot 13, 4596; Chas G Corkey, own, 1573 E 23rd (So 5210); Guy V Colf, bldr, 113 S Larchm Blvd (Holly 8080); $5000 and $300 $5

28,879-80—Dwlg. 2 sto, 9 rm, 35x28, comp tile rf; garage, 18x20, comp rf; 1926 Victo Ave; lot 5, blk 1, Tr 2072; Guy V Colf, o & bldr, 113 S Larchmont Blvd (Holly 808 $7000 and $200 $7

28,884—Dwlg. 4 rms, 24x30; 2757 Coolidge lot 287, Tr 4905; Minnie B La Velle, own, W Ave 37 (31912); L H La Velle, bl same $1

Thanksgiving at the Hiner House. Dr. Hiner is seated at the head of the table with Ann to his right in this 1930s gathering of a few close friends and family for the holiday dinner. Hiner was in charge of the music department at the University of California at Los Angeles by this time. His friend John Phillip Sousa always stayed in the guesthouse, know as Sousa's Nook whenever he came to the city. Troy Evans collection

Kansas City, remembered that "as a kid we were ready at any time to walk a mile to hear 'Doc' Hiner's band play at warm summer evening concerts in public parks." Shipley mentioned that as "A famous soprano on her own, Anna Hiner performed as the favored soloist of John Phillip Sousa, the famed military composer of such patriotic tunes as 'Stars and Stripes Forever'. Sousa was a great friend of the couple."[6]

"When I bought this place nine years ago it was a public dump," Doc Hiner related in his 1929 interview with the *Los Angeles Times*, "and it cost mighty little to make it what it is. Just work and love and enthusiasm. I couldn't saw a board when we started. I hadn't an idea of how to mix concrete or lay cement. I've learned so that now my boys (his students) learn music and manual training together. "

This 1922 photo on the left shows the home being built of wood framing and tells us more: casually hanging off the roof and walls, three of his "boys" and two men take a break from their work to pose for the photographer. The man on the right looked like the lead carpenter with his dark bib overalls, probably complete with pencil pockets and loops on the side for hammers, while on the ground a uniformed man with a shiny belt buckle, dark tie, and tucked in, pressed shirt stands with his arms flat to his side as if ready for an Army inspection. The other folks were identified as "Dr. Hiner's stepson Everett Moore, Anna Hiner and Dr. Hiner."[7]

Ten years later this Thanksgiving dinner photo was taken showing their popularity and family. At the head of the long table sits Doc Hiner appearing ready to direct not a band but a Thanksgiving feast for seventeen of their friends. Dr. Hiner looks over his right shoulder with the bearing of a commanding perfectionist, gained from decades of teaching music and directing bands. His appearance is typical of a professional for that 1930s era with slick hair combed straight, wire rimmed glasses, thin mustache, and a bland commanding stare. In the background stand the cooks with a server proudly holding the main course. In the foreground Anna Hiner matches that style of controlled devotion to detail in this formal moment looking absent of humor and conviviality. And these characteristics are reflected in the old English style home with its steep roofs needed in snow country but not for Southern California's sun. The Hiner property was a setting unlike his neighbor's, a bohemian journalist with a flair for promotion and a love for the native Indians, the Missions of California, and a few wives. We visit that home next.

"Any fool can write a book and most of them are doing it; but it takes brains to build a house."
Charles Fletcher Lummis

CHARLES LUMMIS AND EL ALISAL
200 E. AVENUE 43

In the 1890s a young man of incomparable vigor and future accomplishments walked from Chicago to Los Angeles, a distance of 3,500 miles. Without fear of being hit by cars or being robbed, he safely enjoyed his walkabout and sent articles of his journeys to the *Los Angeles Times*. This road side safety says a lot about that era and the lengthy journey displays energy that remained

Charles Lummis' trademark until a stroke years later, after which he recovered much of his functioning. Many books and articles tell us of this remarkable man who relocated to and promoted the developing Southern California of the early twentieth century. A brief review of Lummis' life presents a man's love of his new land and his love of his arroyo stone castle.

Charles Lummis began as a journalist with the *Los Angeles Times* and edited a magazine promoting California, called *Land of Sunshine*, writing "more than 250 pieces for the magazine."[8] In his spare time, he was the L.A. city librarian for a few years and then founded the Southwest Museum, which focused on the culture of the Southwest Indians, whom he admired intellectually. An artistic workaholic whose twenty-hour work day routine began with writing in the morning, then building his stone castle in the afternoon, and then writing for hours at night. "To him there was no day of resting during the year; and four hours sleep at most, he was up early, working at top tension all day long, and nearly all though the night."[9]

Lummis' diary entry of 1906 tells of this workaholism, "6 hours at the carpenter's bench which resulted in my putting the grille in Bertha's door and hanging it on its hinges."[10]

This lifestyle is not recommended. His personal life lacked physical health (he had a stroke, taking five years to recuperate) and domestic health (he divorced his wife after 12 years of marriage while enjoying many other sexual partners). Mark Thompson, in his book on Lummis, *An*

When Los Angeles City acquired El Alisal vines and shrubs had been left unattended for years, yet the home's solid beauty remained.

Lummis' living room filled with Native American artifacts.

American Character, describes the building of El Alisal: "In 1894, Harvard educated Lummis began seeking a new homesite and the area he focused his search...the rugged Arroyo Seco, a rocky watercourse, extending for more than 10 miles from Los Angeles to Pasadena...was filled with the scattered homesteads of artists, writers and assorted self-styled bohemian intellectuals."[11] Lummis fit right in.

In May 1894 he found a two-and-a-half acre lot for $650 and built El Alisal. In 1895 Lummis purchased multiple adjacent lots in the lower area of the Arroyo Seco. In his diary he referred to this area as “Sycamore Grove”. Here he constructed his stone castle with the help of his young son and Mexican groundskeeper. Lummis named it ‘El Alisal’, the Sycamores. “he built his house,” writes Charles Moore, Peter Becker, and Regula Campbell in their book on Los Angeles architecture, “with his own hands between 1897 and 1910 as an example of architecture and the way of life he had spent most of his life crusading for.”[12] The materials used to build his home Lummis said, “would last for a thousand years.”

Its unique design included a crenellated circular tower, mission style arched parapets, and, to include respect for the native Indians of the area, he placed a metate grinding stone near the base of the tower. El Alisal evolved into a large two story of thirteen rooms and four wings framing a central courtyard that became the patio hosting many parties. Rough hewn timbers, wood plank floors, and plastered walls where Navajo rugs hung completed the interior.

“The lot had several features that Lummis liked,” Thompson wrote, “including a towering grove of old sycamore trees and a seemingly limitless quantity of large water-worn boulders out of which he intended to build a castle for the ages in the shade of the sycamores. He had been

Charles Fletcher Lummis, the Harvard educated grandee of the Arroyo Seco and author of the magazine* Land of Sunshine *and founder of the Southwest Museum.
Photograph from www.seasonalchef.com, biography of C.F. Lummis

mulling over his dream home for a decade. Using an old California-Spanish word for sycamore grove, he christened his home El Alisal. Practically everyday that he was in town for the next 14 years (another source says it took 17 years to build) Lummis spent at least a few hours at hard labor hauling rocks and mortar...he spent hours each day setting boulders into place, pouring cement, and hewing beams and planks for floors, ceilings, door and window frames, in addition to his eight or ten or twelve hours spent on magazine (the Southwest) and association work."[13]

"His gym," he called it. A way to get daily exercise. And lifting 30-60 pound stones definitely provided a good workout.

Historian Kevin Starr gave a perfect description of the colorful idiosyncratic man describing him "styling himself as 'Don Carlos' (the king of Spain in the sixteenth century), Lummis lived on Southwestern food—chili, tamales, frijoles, olives—and wore a green corduroy suit cut in the Spanish style, set off by boots, a brightly colored cummerbund, a frilled shirt, a broad brimmed hat and Navaho (sp) jewelry."

Then in 1928, he wrote a foreboding diary entry, "Dr. Moland went over to me. I said: 'It got away from us, hasn't it—and gone much farther than you expected.' He said: 'Yes, it has—I can't quite understand it.'

"Anyhow, it's in damned bad shape.

"Miss Christe applied the radium and I went to sleep on it for an hour and a half.'"[14] Charles Lummis left his one-of-a-kind stone castle, unfinished, dying on November 25, 1928.

Lummis was a modest sized man, weighing maybe 150 pounds judging by his pictures, in his forties, who working with his son (who probably mixed the mortar) and the Mexican gardener, moved granite boulders weighing thirty to a hundred pounds up the sloping bank of the wash a quarter-mile distance to his building site, then put them into their grouted places in a two-story high wall,' a surprising feat his slight frame achieved. With his stone home almost finished he realized his dream of living and partying in a medieval stone castle with its Spanish flair.

His home is now managed by the Los Angeles Parks and Recreation Department and can be visited at 200 E. Avenue 43, Highland Park, while our next 'bohemian' home remains a private residence owned by builder's family.

THE ABBEY SAN ENCINO

6211 ARROYO GLEN

Possibly enticed by his friend Charles Lummis, Clyde Browne built his rock home and print shop in the Arroyo Seco canyon, not to be outdone in realizing dreams, nor to be outdistanced in the time taken to construct his rock dream.

Born in Ohio in 1872, Browne arrived to Los Angeles in 1904 and worked at the *Los Angeles Examiner*. A few years later he quit to chase his dream of publishing fine books, and to achieve this he needed a home and print shop. For ten years he collected stones and artifacts and participated in arduous construction. By 1915 the Abbey was completed, and two years later Los Angeles City, under the engineering skills of Mulholland, finished the 250 mile aqueduct from Owens Valley, bringing water to the desert giving rise to exponential growth and the building of many cobblestone homes.

This stained glass depicts a Franciscan printer along with an American Indian boy working a hand press.

Clyde Browne's new 2,600 square foot Abbey San Encino was named after the native oak trees and incorporated dozens of artifacts in its construction—crucifixes, lanterns, stained glass windows, and more many obtained by his friends on their European travels. The Abbey had a large cellar, a jail, a bell tower, and the print shop in the design form of a chapel. Moore noted in his book *Los Angeles* "In the Spanish and Mexican tradition, the house wraps around a large open patio... the walls blocks of hand-carved sandstone and granite and round fieldstones" Moore found the Abbey theatrical and its various forms and artifacts "make this rancho seem immediately familiar; it is a real part of California's make-believe past."[15]

Browne raised an artistic family and became the grandfather of singer Jackson Browne and song writer Severin Browne, who continue to live here three generations later.

For centuries canyons have attracted artists, possibly to avoid the distractions of city life, or to avoid people so as to focus on their inner muse's inspirations. Clyde Browne dreamed of creating an artists community and built his home and workshops for others to use. Floyd Bariscale wrote, "Inspired by Elbert Hubbard's Roycrofters, Browne moved a garage on his land and, in the summer of 1915, began the decade-long project of building his home, Abbey San Encino. Browne modeled the home on a variety of places, like the Mary Queen of Scots Chapel at Holyrood Castle, the San Francisco de Solano Mission at Sonoma, and Carmel's San Carlos Borromeo Mission."[16] Stones from the nearby Arroyo Seco wash were brought up on a narrow gauge railroad Clyde Browne built.

Artifacts collected from America and Europe included the big, round stained-glass window on the building's south side. This stained glass shows to all that this was a printer's workshop for it depicts a Franciscan printer along with an American Indian boy working a hand press. In 1972 Los Angeles declared The Abbey historical landmark #106, which can be viewed on YouTube.

811 BRIDEWELL STREET

A freeway now extends down the bottom of the Arroyo Seco Canyon connecting Pasadena to downtown Los Angeles. The stone home at 811 Bridewell looks across a narrow street to that freeway, about fifty feet distant. But in 1924 the masons building this magnificent bungalow would have been next to a stream shaded by sycamore trees, a bucolic setting of quiet and peace while being convenient to downtown Los Angeles.

For $5,000 Albert P. Orecchia built this home, which included two bedrooms and one bathroom within its 1,370 square feet, providing spacious rooms for his family. This neighborhood had been subdivided into modest residential lots, with electricity and water being provided by Los Angeles. In late 2019 this home was listed for

Without bracing, the rock chimney remains undamaged from a century of earthquakes. Though taller ones have toppled, the height of this and others tells much of their strength.

811 Bridewell, Los Angeles, back of home and interior.

$800,000, an unheard of price in 1924. Since then a small apartment building was constructed next door indicating that the city rezoned this area to multifamily buildings threatening that this stone home may be demolished. Since rezoning, the city has not listed this unique home under the 'historic preservation review' status. Thus nothing is protecting it from strange renovations or demolition. A recent drive-by showed that, fortunately, the new owners show no intention of destroying this building.

The masonry skills shown in this building appear in the large boulders above the windows and doors, the narrow mortar joints and the sensitive interlocking of the stones. All hallmarks of a master mason. This was constructed during times when a similar sized wood-framed house could be built for a couple thousand less.

END NOTES

1 Weaver, J., *Los Angeles The Erroneous Village.*, Capra Press, 1980, p. 51-2.

2 Ibid., p. 55.

3 Ibid., p. 99.

4 Ibid., p. 67.

5 Meares, H., “The House That Love and Music Built, the Hiner Home of Highland Park,” KCET Link, September 26, 2014.

6 Ibid.

7 Fisher, C.J., *Images of America, Highland Park*, Arcadia Publishing, 2005.

8 Thompson, M., *American Character: The Curious Life of Charles Fletcher Lummis*, Arcade Publishing, 2001, p. 85.

9 Lummis, C.F., *My Friend Will*, Leopold Library Classic, 2nd Ed., 1912, p. 16.

10 Kipen, D., *Dear Los Angeles: The City in Diaries and Letters, 1542 to 2018*, Modern Library, 2018, p. 372.

11 Thompson, M., *American Character: The Curious Life of Charles Fletcher Lummis*, Arcade Publishing, 2001, p. 85.

12 Moore, C., et al, *The City Observed, Los Angeles: A Guid to Its Architecture and Landscapes*, Hennessey and Ingalls, 1998, p. 66.

13 Thompson, M., *American Character: The Curious Life of Charles Fletcher Lummis,* Arcade Publishing, 2001, p. 191.

14 Kipen, D., *Dear Los Angeles: The City in Diaries and Letters, 1542 to 2018*, Modern Library, 2018, p. 238.

15 Moore, C., et al, *The City Observed, Los Angeles: A Guid to Its Architecture and Landscapes*, Hennessey and Ingalls, 1998, p. 66.

16 Bariscale, F.B., “Abbey San Encino, no. 106” bigorangelandmark.blogspot.com, January 17, 2008.

The Arroyo Seco after a flood, possibly the 1934 deluge that destroyed many homes in the foothills, showing the collapsing of the railroad bridge.

Here, L.A.'s response to SoCal's infrequent but torrential rains awakened an engineering response. The Arroyo River was contained in concrete channels for its entire length making the adjacent lands more livable and profitable, but destroyed the natural recharging to the ground water, and in this 'Cadillac Desert' that exasperates SoCal's ground water supply especially needed in years of drought.

CHAPTER 7

Glendale, Burbank, and an Earthquake Discussion

On our Way we met the entire population of an Indian village engaged in harvesting seeds on the (Los Angeles) plain. In the afternoon there were other earthquakes; the frequency of them amazed us.

Miguel Costano, 1769[1]

While admiring the sunset over the Pacific Ocean where orange and red clouds framed the sky in 1928, Ed Weston with his photographer's eye wrote, "I realized how much one misses living in an ugly, squat city like Glendale, surrounded by box-like abominations in brick, or worse, stucco and cardboard pretensions."[2] He definitely did not see the river-rock beauties on Olive Street.

These homes provide us with a hint of the character common to their owners and masons. They suggest independent thinkers, a solid dependable constitution, and someone planning to stay. A cobblestone home may have also been seen as a strong statement of one's status or wealth. Like Orlando C. Lane, owner and founder of Burbank's first Ford dealership.

902 EAST OLIVE STREET, BURBANK

It may have been on his forty-second birthday in 1923 when Lane moved his family into their new cobblestone home on 8th and Olive Streets. Unlike the Arroyo Seco culture of artists and writers, Orlando Lane enjoyed the rewards of being a successful businessman. In his youth Lane displayed his mechanical and business abilities by opening a bicycle repair shop. At the turn of the twentieth century bicycling grew in popularity both in cities and suburbs which provided Lane with the opportunity to capitalize on this trend. A few years later he opened an auto repair shop, undoubtedly one of the first in the Burbank area, and in 1912, at age thirty-three,"finally a Ford auto dealership."[3] His success allowed him to sell the dealership and retire in his early forties, possibly using some of the proceeds to build their new home.

Expanding from a bicycle repair shop to a new car

902 East Olive Street, Burbank

On the side of the garage is a stone boulder that I calculated to weigh about 224 pounds. And the larger boulder below it, the one with the black spot on its face, weighs about 302 pounds. How were these sized boulders placed six to seven feet above ground in the days before front loaders and forklifts? Men of super strength, strength beyond our office capacity of today, built these stone buildings.

dealership reveals excellent business acumen, as does being able to retire in only fifteen years. Opening a dealership requires capital which Lane either saved or found in a rich partner.

In a 2010 *Los Angeles Times* article by Joyce Rudolph the present owner mentions that "'The Rock House' is a Historic Structure of Merit because of the river rock architecture." This article also mentions that "the house is one of only a handful of buildings in the city that uses that unique river rock."[4] Its neighbor is also a home of distinction.

906 EAST OLIVE

Built in 1922, and possibly using the same mason as Orlando Lane, this two-story home sheltered a family, as its four bedrooms and two bathrooms suggests. This area of Burbank attracted upper middle class families and the size of the home along with its detailed masonry, as seen in the tight mortar joints and skilled placement of the interlocking stones, shows that these homes were built by moneyed people.

DEUKMEJIAN WILDERNESS PARK AND LE MESNAGER BARN

George Le Mesnager and Pierre Dunsmore, immigrants from France, purchased in 1886 approximately 790 acres of foothill land to plant vineyards. In 1914, the French wine makers and businessmen built a large stone barn to process their grapes and age the wine in wooden barrels. But they built in a perilous location, at the bottom of Cook's and Dunsmore Canyons in the San Gabriel Mountains. During the 1933–1934 winter rains, the same rains that destroyed much of La Cresenta, these two canyons discharged tidal waves of mud through the vineyards destroying the wine and vines but not the stone barn or Le Mesnager and Dunsmore's resilience. A month earlier in November, a massive forest fire burned the local foothills and destroyed the barn's roof, wooden parts, and contents. Damaged but undeterred the Le Mesnager family rebuilt in 1934 during the Depression and then replanted the vineyards in 1937 to restart wine production.

The Le Mesnager family occupied these acres from 1890 to 1960 then sold the land to a developer. When the developer could not get the necessary permits, the City of Glendale purchased the land in 1988 and restored the barn and land into Deukmejian Wilderness Park. Glendale also listed this stone building as a historic building noting that "The stone barn is of historical significance as a rare example of a two-story vernacular rock structure."[5]

Most masons who built the stone buildings remain unknown, yet we are lucky here for we know of Pierre Escalle, "another French immigrant who was both a stone mason and vintner, lived in the adjacent community of La Cresenta" and would have likely assisted in its construction. Many stone masons had emigrated from France in late 1880s and Escalle, born in 1854 in St. Eusele near the French Alps, knew his French neighbors, the Le Mesnager family. A close friendship probably occurred where Escalle helped with the stone work on the

The stone barn restored. Note the dark squares. That's where the wood beams cantilevered from the wall to support the scaffolding needed to build the walls.

The handwritten text at the top of the photo says "George Le Mesnager 1919–returning from WWI. He had bought the bare land in La Cresenta, now Glendale in 1885 and yet he returned to join the French military during WWI. Le Mesnager showing his loyalty to his country to became the oldest combatant in the French Army at 64 years old. At the end of WWI he turned 69 with many medals awarded for valor while in combat."

The Le Mesnager barn survived Prohibition, fires, floods, and earthquakes, but nearly did not survive a developer's wrecking ball in the 1980s. Fortunately the City of Glendale purchased the 700-acre site for use as a park and bolted steel beams to sides of the barn to stabilize the structure seismically. The beams will be removed as the building undergoes a more complex seismic retrofit by the city in coming years. (Courtesy Glendale Library.)

The interior of the barn once had a second story living area that was used by the son and grandson of George Le Mesnager from the 1930s until its sale in the late 1960s. After that, the barn was used to stable horses. The City of Glendale acquired the barn as the centerpiece of the Deukmajian Wilderness Park and plans to refurbish the historic structure for use as a community center. (Courtesy Glendale Library.)

barn and the elder Mesnager gave him advice on growing vineyards and wine making. Escalle also built the now-demolished Gould Castle, located high in the foothills, listed in Rockitecture Book II in chapter 13A.

The construction of the large stone barn in 1915 and the expansion of the vineyards to 1,300 acres shows the profitability of being a vintner before Prohibition forced Le Mesnager to switch to selling the grapes and producing only religious wine.

Hardship continued ten years later when the forest fires of 1933 destroyed the barn's wood-framed roof and its hundreds of barrels of wine; then a month later, intense rains eroded the barren hillsides and mud flows destroyed the vineyards. Most of the San Gabriel foothill suffered large building losses from this 1934 flood. In the following years, the Army Corps of Engineers constructed concrete debris dams and flood control channels which protected many cities in the 1938 floods and opened more lands for development.

In February, 1933 Anton Wagner took this photo of adjacent Burbank land, showing typical condition of undeveloped land that Le Mesnager would have bought in 1885.

Today, volunteers, under the guidance of Stuart Byles from La Cresenta, plant, harvest, and maintain the small vineyards on the south side of the stone barn.

Constructing with field stones in the hills of France was familiar to Le Mesnager, and his stone barn in the San Gabriel Mountain foothills carries on that traditional masonry skill.

In chapter one, we met Stuart Byles at the St. Luke's of the Mountains Church. He also volunteers to maintain the small vineyard of Alicante grapes next to the stone barn that Le Mesnager owned and operated. The pressing of the annual grape crop, the same varieties as the originals, produces about one hundred bottles of wine, which are shared by the all volunteers assisting him. With Glendale's far-sighted renewal of the two-story stone building and Byles' efforts in regenerating the wine production, Southern California's French community, stone buildings, and wine heritage continue to live.

Le Mesnager, a man loyal to both his vineyards and his native country, left Glendale and returned to France to fight as a soldier in the French army during WWI. His dedication did not stop there for after the war he once again returned to France, not once but three times in seven years, to help his country's reconstruction. The journeys occurred by ship and, when he turned sixty-nine in 1919, he become known as the oldest enlisted soldier in France's army. A few years later, in 1923, he died in at his old hometown of Mayene.

ADDITIONAL STONE HOMES IN GLENDALE AND BURBANK

1515 W. Kenneth Road., Glendale

Plaque at the Historic Gates to the Sunset Canyon Country Club of the 1920s. The gates, restored by Burbank City, are located at East Olive and Sunset Canyon Drive.

These stone homes were possibly built in the 1920-1930s on Country Club Drive, Burbank, in the Verdugo Mountains framing the south side of La Cresenta Valley

Two homes at 906 Country Club Drive using canyon rock construction for the first floor and garages. The original rock homes were built in the 1920s, while the second floor additions occurred after WWII.

On the left we found a small cobblestone cottage hidden by a new wall along a driveway below 800 Country Club. The original one-story home hosted a wood second floor addition, which probably occurred in the 1950s.

EARTHQUAKES, CLAY BRICKS, GRANITE STONES, AND MORTAR

"Always a factor in California, earthquakes are something every Californian must live with but few care to think about..."[6]

"Earthquakes are not pleasant, to be sure, but the reality is hardly so dreadful as that...Severe earthquakes are not unknown in California: witness the shock that destroyed the mission church in Capistrano in 1812...and the disaster in 1872 at Lone Pine; and what has occurred here may occur again at any hour or minute."[7]

3722 4th Street, La Crescenta

The rolling shaking ground must have startled the cows, horses, and especially the eastern settlers from where the ground rarely moved. But here in California it moved often, and still does. This section may seem confusing, but no geologist will ever tell you that earthquakes are simple. Yet, all of you who have experienced an earthquake know that the violent, quick earth movement tosses objects around that crash in different places, and often break. This happens not only to books, dishes, and furniture but also to buildings and bridges.

In March 1933, in the early evening, a 6.4 earthquake destroyed the City of Long Beach and severely damaged buildings in the surrounding communities. Approximately one hundred and twenty people were killed by falling bricks and debris as they ran out of buildings to avoid the chaos inside of ceilings falling and goods being tossed about.

In 1860, Ralph Waldo Emerson understood the essence of geology when he wrote, "we learn geology

The Crescenta Valley butts up against the mostly undeveloped San Gabriel and Verdugo mountains. As such, residents get to enjoy the incredible beauty of the surroundings, until wind, earthquakes, fires, and floods shatter the tranquility. Shown here is the devastated Le Mesnager barn and vineyards in 1934—a double victim. A fire in November 1933 burned the structure and grapevines and a flood in January 1934 stripped off the topsoil. In this photograph, one can clearly see the lighter colored path of the flood that came out of Dunsmore Canyon and swept straight across the valley toward the Verdugos. It was this flood channel that damaged the back portion of the

Top left: ***Woosley fire in Malibu Hills, California, started on November 8, 2018 and burned 97,000 acres.***

the morning after the earthquake." Our stout stone buildings survived this Long Beach event while many brick and wooden buildings collapsed. They also survived the 7.3 Tehacapi quake in 1952. Then the 6.6 Sylmar quake on February 9 at six a.m. in 1971 that killed 64 people and "ripped along a twelve mile fault beneath the San Gabriel Mountains."[8] This Sylmar seismic event crumpled the Olive View Hospital built just a few months earlier and was yet to be occupied, and freeway overpasses built to prevailing codes for earthquake safety collapsed. Though buildings, bridges, and roads suffered significant damage all over Southern California from these seismic events, the owners and occupants of century-old river rock buildings told me that their homes seldom had a crack from the same earthquakes.

In studying these cobblestone buildings a curious person might ask, "How come? Why didn't they crack or fall apart?" The craft of building with cobblestones can become the science of earthquake-resistant design. Neanderthal engineers and egos aside, when a building does not collapse during a seismic event it demands studying. We depend on the veracity of building codes for our safety, and they need to be based on science not conjecture or

Olive View Hospital after the Sylmar earthquake of 1971. The flanking stair towers fell away from the main building, along with other severe damage. The new building—a few months old and fortunately, not yet occupied—was demolished.

waiting until a tragedy occurs. Our lives cannot be lab experiments. This lack of research and understanding of earthquake design, not only in calculations but also in shape of the structure, could motivate engineers and architects and city official to search for more answers using rigorous standards, not opinions or conjecture or preconceptions. Study traditional designs such as these buildings and Japanese structures as a whole, not just bits and pieces like beams or columns or joints, but whole units. Once the whole structure is understood, then the individual parts can be reviewed as needed. Along with more thorough understanding, using indigenous materials is the essence of energy savings.

In 1991, along the Sierra Madre fault, which extends along the southerly edge of the San Gabriel Mountains, a modest 5.6 quake occurred in the morning waking many. $30-$40 million of damage occurred to hundreds of homes and apartment buildings. Dozens of buildings in the Glendale-Pasadena area were 'red-tagged', deemed uninhabitable by the city's building departments, but the river rock buildings in the canyons of Sierra

Madre and those in the nearby communities of La Cresenta, Pasadena, San Fernando Valley, or Claremont/ Upland remained undamaged.

Mother Earth, not done shifting and shaking Southern California, shook the region again in 1992 and again in 1994. The 1994 Northridge 6.9 earthquake resulted in significant property damage, around $44 billion. And still the century-old river rock buildings of the nearby communities, such as historic Stonehurst community of sixty-four stone homes, and dozens more stone homes in San Fernando City and La Cresenta Valley remained undamaged. Even their home's windows did not break, though the occupants needed a while to calm down and straighten up fallen objects inside.

Since then, SoCal's many fault systems have remained quiet, probably exercising at its geological gym gaining strength for another big quake.

To understand the strength of buildings in resisting earthquakes, a quick review of geometry and its application to construction is helpful. The shape of the building, along with what materials are used, can strengthen or weaken the building. Though we build in rectangular shapes, it is the circular form that is more resilient to earthquakes and mud flows.

Stones' irregularity may add significant strength to the building.

The Indians of California knew the gods of the ground were unsettled and built accordingly, with branches and fronds in conical shapes, light weight and stable bee hive structures. (Recreations can be visited at the Palm Canyon

Reservation in Palm Springs and the former Chumash site at Satwiwa past Newberry Park south of the 101 freeway). Rectangular structures with their ninety-degree corners and the interior junctions of the walls and roof are inherently weak geometry. Straight lines deflect under pressure. Corners pull apart and separate. Curved or arched lines maintain their shape under the same pressure as there are no corners. Press a hula-hoop for example. It resists. Now press the side of a wire coat hanger or box; the side deflects. Inherent in the shape of the arch is that it resists forces and deformation by compression. This is seen in stone arches above windows and doors. The stones do no fall because the shape resolves the compressive forces; the stones push on each other tightening them, strengthening the structure.

In 1933 Long Beach collapsed.

Build like a fishermen's boat and your home will ride out the ground waves like a yacht riding the ocean waves, undamaged. In all these quakes the destruction occurred in wood-framed and brick construction, not to the buildings of granite stone construction.

Susan Straight, in her newspaper article on the cobblestone homes in Upland, interviewed owners Randy and Susan Owens who live "on Sixth, in a stone house with a distinctive chimney built in 1937. They bought the house in 1966 and, when asked about earthquakes they laugh. 'This place holds up pretty well during an earthquake...We've felt plenty of them. I don"t worry,' Susan Owens said."[9]

Possibly the regular rows and the straight lines in brick construction contributed to the weakness of brick buildings. And maybe the relative softness of the fired clay of bricks also is a weakness compared to granite. In river rock construction all the stones are different sizes and straight lines do not occur. The stones are natural unshaped granite, a hard rock with crystals geologists name "white quartz and plagioclase, pink feldspar, and black biotite and hornblende."[10] These different minerals give granite stones their sparkle and salt and pepper look. Granite is the hardest material in nature, second only to diamonds. As you may remember, subterranean volcanoes made granite; and then, over millions of years, seismic forces compressed the cooled stone masses.

Of equal importance to the rocks themselves is the strength of the mortar binding the stones. In the late nineteenth century Portland cement became available which provided a significantly stronger bond and weathering resistance compared to the common lime mortar used up to that time. We will skip over the details of the chemical reactions when the mortar dries and sets, as for us the important qualities are its strength and resistance to erosion. When the mason buys the Portland cement he receives a 50 pound bag of fine powder that is caustic and burns the skin. But after adding water and mixing it into a mud-like consistency, complex chemical reactions begin and drying continues for months; this is the curing stage when strength increases with the slow crystallization and "interlocking of the crystals (that) give cement its strength."[11]

A typical brick building damaged in the 1933 Long Beach earthquake.

Looking at the buildings damaged by the Long Beach earthquake you will notice piles of individual bricks. Why individual? The mortar in the joints failed. These buildings were constructed with sand and lime mortar joints, crumbly material that your finger can scrap out like sand. Portland cement, however, though not used than in many brick buildings either to save money or through ignorance, added a critical bonding agent, resulting in strong non-erosive joints. And Portland cement is the mortar we found in all the cobblestone buildings, and it has remained dense and firm even after a hundred years.

Another interesting fact: bricks are fired at high temperatures in kilns to hardened and bake the clay which evaporates all its moisture. Granite cobblestones come from washes, the ground, and river beds, all wet areas leaving the stones with residual moisture from sitting in moist conditions for thousands of years. Why is this important? Mortar uses water to mix its products of sand, cement, and lime together into a workable product the consistency of thick mashed potatoes. To get a strong bond between the stones or between the bricks requires a thorough connection with the mortar. It can't be just applied like dry peanut butter on toast which easily falls off. Moisture is required in both the cement mortar and the brick or stone to avoid a dry bond, a bond not like paint on glass, but a bond like strong penetrating glue. The dry brick sucks up the moisture drying out the mortar which then results in a dry, weak bond that fails with movement. Just think of beach sand: wet sand binds and stays together; beach sculptures are made out of wet sand. Dry sand? The best shape you may get is the shape of a cone or mound.

But we must consider another idea. Earthquakes make waves in the land that "travel at speeds of more than a mile a second through the Earth."[12] That's 360 miles per hour. These waves are called seismic waves and act like the waves of the ocean created by winds. Boats are designed to float and move, rock up and down and side to side; houses are designed to stay put. Well, that's the intent. Boats, when at rest moored in the harbor or marina, bob up and down riding the peaks and troughs of small waves. These monolithic cobblestone homes ride the land waves generated by earthquakes.

Richard, who owns and lives in a two-story cobblestone home on the south side of La Cresenta Valley, shared his experience while gardening in the early morning. "I am on my knees tending to some plants when it occurred. The land shock, it rolled, and I heard that deep rumble." Knowing what was occurring he looked over his shoulder to see his cobblestone home rolling and riding the ground waves like a boat in the ocean. "No cracks, no broken glass, no stones thrown on the ground, things were pretty tossed around inside though."

END NOTES

1 Kipen, D., *Dear Los Angeles: The City in Diaries and Letters, 1542 to 2018*, Modern Library, 2018, p. 271.

2 Ibid., p. 44.

3 "Rock House, 902 East Olive Avenue," Burbank Historic Resources.

4. Ibid.

5 "Deukmejan Wilderness Park," www.glendaleca.gov/deukmejanwildernesspark, accessed October 10, 2015.

6 Hayes, D., *Historical Atlas of California*, University of California Press, 2007, p. 233.

7 Carter, C.F., *Some By-Ways of California*, The Grafton Press, 1911, p. 160.

8 Martholomew, D., "Sylmar-San Fernando Earthquake: 45 Years Ago Tuesday," Los Angeles Daily News, February 8, 2016.

9 Straight, S., "Rock Walls and Stone Sunbursts: The Quarry-faced Homes of Southern California's Past," KCET, July 26, 2012.

10 Dvorak, J. *Earthquake Storms: The Fascinating History and Volatile Future of the San Andreas Fault*, Pegasus Books, 2014.

11 "Portland Cement," Wikipedia, https://en.wikipedia.org/wiki/Portland_cement, accessed June 6, 2020.

12 Dvorak, J., *Earthquake Storms: The Fascinating History and Volatile Future of the San Andreas Fault*, Pegasus Books, 2014.

CHAPTER 8

South Pasadena, Pasadena, Nick the Greek, and Some Technical Thoughts

For my own taste I prefer Pasadena, the "Crown of the Valley"...At the foot of the Sierra Madre range...Pasadena seems to me to be as near Eden as can be found by mortal man.

Kate Sanborn, 1906

PASADENA

Pasadena's main street, Colorado Boulevard, began as a dirt road for pedestrians and carriages; cars, pavement, and smog would be sixty years in the future. Women wore ankle length dresses while the men wore dark wool suits and wool hats as they strolled down the dirt road.

The first buildings constructed in this new town were made of wood so the proprietor could open his store quickly. Notice the "carpenter" and "lumber" yard signs, proof that wood framing was the preferred building method. Hiram Reed writes in 1895 that "The first lumber yards started in 1883 on the north side of Colorado Street,"[1] and that was possibly the lumber yard shown in this undated historical photo. A few years later, a couple river rock buildings would be built on the east side of Pasadena, out in orchard land by the dry wash, "out in the sticks," the townspeople said. And in 1890 the Tournament of Roses started as an advertising campaign by the Valley Hunt Club of Pasadena to entice tourists to visit the sun-filled land.[2]

Colorado Boulevard, Pasadena in the late 1800s.

A century ago tourists of eastern industrial cities visited Pasadena (then named 'Indiana Colony') to escape their winter months of blizzards and cold gray skies. Throughout its 135 year history, Pasadena has been visited and settled by well-heeled people, who wanted to partake of its year-round sun, mild temperatures, and intellectual and artistic culture at the nearby universities, museums, and theaters. The story of creating Pasadena on sage-covered alluvial fans into at first a vibrant land of vineyards and citrus groves and then into a vibrant city of tens of thousands of residents illustrates the founding and fast growth of these San Gabriel Mountain foothill towns. Man's basic desires formed all these towns—the search for wealth and health. We all seek a pleasant and pain-free life.

Before the 1700s, the government of Mexico issued land grants "to individuals that made application. And the northwest section of Rancho San Pasqual became Pasadena and South Pasadena.[3] These communities grew in popularity, for in the mid-1880s as "Pasadena contended with Los Angeles as an attraction for an outing, for dominance of the Southern California basin. Both cities were ceasing to be sleepy places at the far end of nowhere. Both were moving ahead and growing to resemble the towns new immigrants had left behind."[4]

Cars, electric trolleys, bicycles, and pedestrians all occupied city streets in the early twentieth century.

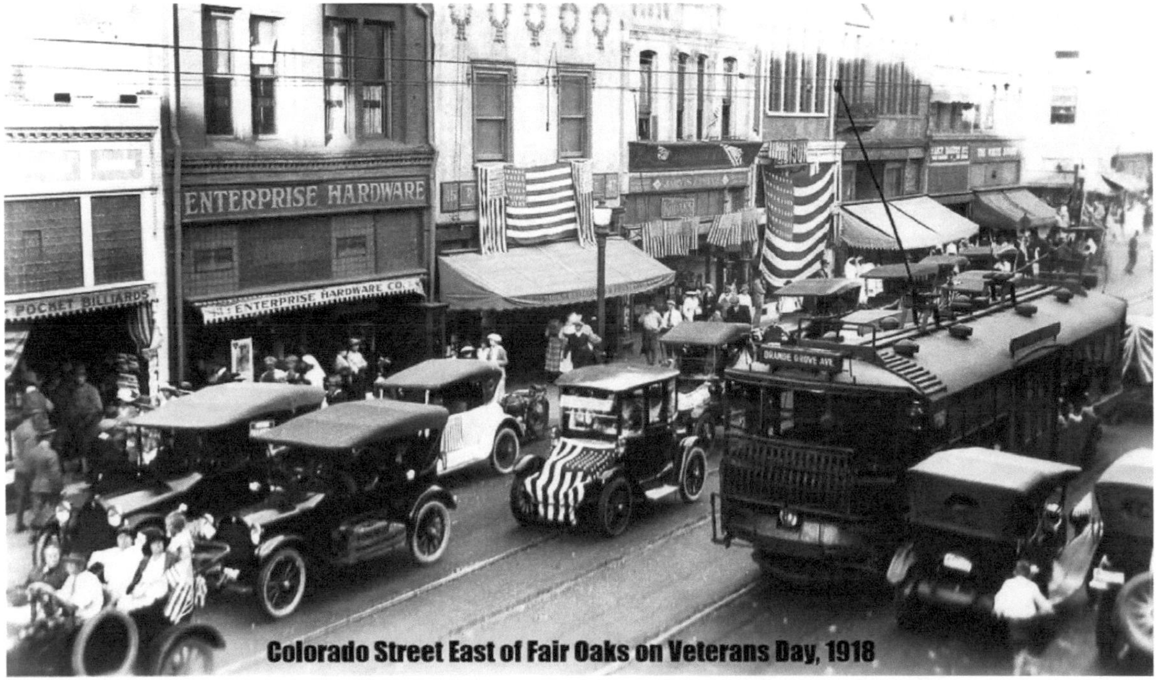

Wealthy industrialists of the mideast preferred Pasadena and Altadena to build their winter estates and retirement mansions. In 1880, E. D. Holton, a former U.S. Army soldier wrote of his visit here, "Never before...have I seen any such sunlight, transparency and luminousness of atmosphere as characterize the days here in Southern California."

Pasadena "was started upon virgin land by immigrants from Indiana (thus the name 'Indiana Colony')," wrote an early settler. "It is situated directly before the Sierra Madre Mountains, upon an open, undulating plain. Today it is a garden spot. Handsome dwellings rise up amid orange and lemon groves,...roads and churches and school houses attest the character of the people. A system of water works, carried by iron pipes, spreads throughout the settlement."[5]

It was said that at one time during the early twentieth century a thousand people a day arrived in Los Angeles to buy a lot and build a home. The weather was a strong reason, as Ms. Sanborn wrote in 1906, "Then the Climate-spell it with a capital..to a tenderfoot straight from New Hampshire, where we have nine months of winter and three of pretty cold weather, this seems like an enchanted ground. A climate warm...Spanish breezes, Italian sky and sunsets, Alpine mountains, a nearly uniform climate."[6]

She also wrote of the downsides of living in Pasadena: an occasional earthquake, the Santa Ana wind that "blows off steeples and tears down fragile structures." Earlier in December of 1891, "the little church (the Methodist Church in Pasadena) was demolished by a wind storm," wrote Reed, A "total wreck."[7]

Reed related how church steeples were "blown off into the streets," roofs torn off and even brick chimneys were blown down, along with skylights and windows being shattered. Buying glass in these years required money plus the effort to ship it in from the east which caused delays of a few weeks.

Rain created more problems. At the end of the nineteenth century averaged 24 inches, and varied from a light touch of 14.2 inches per season to deluges of up to 46.2 inches. On January 2, 1910, the *Los Angeles Herald* reported, "The Tournament of Roses Parade was assembled yesterday...after a pouring rain stopped." Firemen rushed to rescue people stranded in their cars and homes from the downpour "which has lasted several days."

Yet, immigrants continued to arrive and stay, for many reasons. Throughout history people migrated seeking economic opportunity and dreams of a new life. Such was the situation for Nick Papaioanu.

Using rock for his new home may have been natural to him. After all, he grew up in Greece, a land of famous architecture built of stone—beautiful and iconic temples like the Parthenon, so perfect that the Romans copied their Grecian designs for centuries as the essence of great architecture, and the newly born country of the United States must have had an unspoken agreement that all significant public buildings should look like Grecian temples adorned with a stone-clad finish, stone pediments, and fluted stone columns. So, Papaioanu built his new home of stone, as did his neighbors nearer the wash to the east of his lot.

Papaioanu's family home on Sierra Madre Boulevard and Del Rey Avenue.

314 DEL REY AVENUE AND NICK THE GREEK

Formidable, talkative and friendly, Nick Papaioanu Jr. set a casual atmosphere on our first meeting as if we were sitting in his patio sipping ouzos. He told stories which took us back to the early twentieth century and the life of his father, Nick Sr., the Greek immigrant who built this well-preserved rock home.

He showed us the inside of the adjacent cobblestone garage. He lifted the garage door to reveal two modern sports cars sitting quiet and still, their dynamic potential energy waiting to be released like a child being held back from running to the circus. They gave a vivid contrast to the rooted nature of this stone home, like opening an ancient White Owl cigar box and finding two emeralds on a diamond necklace. I had to adjust what century I stood in and stared for a moment. The metaphor between them, the home's timeless silence of stones and today's fast-paced life of swift cars, airplanes' deafening noise, and our instant digital communication mesmerized me.

Earlier I had sent a blind letter to to the Del Rey address and as a proud owner, Nick replied calling to set an appointment. He was around 60, robust, stout, and strong. Nick talked, his wife joining in at times, and I listened for a couple hours. Nick's pride in his heritage shows as the rock home remains in his family since his father built it. Meeting the son of the original builder is rare and a special experience, one that I savored.

Nick spoke volubly and quickly, "Dad built this rock house. Strong and stocky young man, he started constructing August 15, 1925. I don't know how long it took to build the house. He hauled all the rocks up from the river bed before it was cemented over." Located half a block east, the seasonal creek became a concrete flood control channel after the 1938 floods. When Nick Senior built his stone home, Pasadenans called this area 'way out east' and 'in the sticks'.

"He was a strong man. Worked in produce." Nick said, "One day, my dad told me, fellow workers challenged him, betting that he couldn't pick up five sacks of cement. Now each one weighs eighty to one hundred pounds. He took them up on the bet. Stuck a sack under each arm, held one in each hand and picked the fifth sack up by his teeth. The guy was super strong."

Nick Senior came to the U. S. with a cousin in 1918 from Greece; then probably by train, he traveled from New York to Chicago, but he did not like the cold. Nick mentioned that "after divorcing his first wife in 1930, Dad married Mom. They lived here 'til 1948 or so, and he died of lung cancer in 1954."

Nick proudly shared stories without a pause, as if he was reading from a script: "Dad started a produce business driving his fruit cart down South Orange Grove Avenue (then 'millionaires row' and servants would come out to buy produce), ringing his bell, which I still have. By the time he died he had grown his cart business into a produce business where he sold to every independent market in San Bernardino, Redlands, and Colton. He was a very hard worker, up at 2 a.m., off to the market (downtown LA), then he hired some people and had up to six carts one time.

Nick's father as a young man, probably in the 1920s and dressed for a special ceremony.

Nick Senior sits in the front center with a boy in his lap. Is that Nick Junior?

"Born in 1895, his house in Greece was built out of shale, with a dirt floor. Still there. He was living in South Pas (South Pasadena) and asked some guys building a road down there if he could have the rocks. They said sure. So he took those rocks and built a little house in South Pas. We haven't been able to find it. With this house on Del Rey he hauled the rocks up from the river bed and got some Mexicans masons to help him. This house is built like Fort Knox."

I responded, " Your Dad was a strong man who built his house to match."

Nick replied, "In this house there are not pieces of rock, they are solid big rocks, and in the basement also solid rocks. The interior wall framing is for finish on the inside. In the garage the walls are unfinished with just solid rocks all the way up. (The vertical lines of the 2x4 boards indicate Nick may have formed up the entire wall on the inside, then built the rock walls troweling mortar between the back of the rock and the forms resulting in a cement finish acceptable for garages.)

"The basement became a wine cellar during Prohibition," Nick continued. "He made about 600 gallons of wine per year, (The wine press was in the garage.) 200 for himself, 200 for my grandpa and 200 for my uncle (200 gallons equal 800 quarts equals about 850 bottles.). They put 50 gallon barrels down there." During the Prohibition days of 1920s and early 1930s, making 200 gallons per year per person was allowed for personal use. "So they made it about now (September)," Nick mentioned, "and in late October they would crack the first barrel. To test if it was ready, they would unplug it, put a match over the hole and if stayed lit the wine was ready. If it blew out it wasn't ready. I never got any wine then."

While out in the garage, he reflected on this home. "My mom would

Inside the garage a thin layer of mortar covers the stones. The joints of the 2x4 forms are continuous from floor to ceiling.

never sell it and I won't either. It's kinda like living in a cave, it's very quiet, even when the Rose Parade goes by with all the marching bands (the south side of his stone home faces Sierra Madre Boulevard, the Rose Parade route). When we are sleeping in the back bedroom people could be walking by making noises (after the parade) and it's quiet. It's registered with Pasadena as a historical landmark; it is a rarity to have the original family living in the house. He (Nick Senior) was the only one of my family who immigrated to the U.S. My uncle just died at 102, and my grandmother at 107." As I listened to Nick, I realized that his grandmother lived longer than the stone home.

California is a young state, joining the U.S. in 1850. This family home was built in 1925, and for around 93 years it was never sold, continuously lived in by father and then son, and he hopes his son. Nick continued, "Since 1923 this home has seen the raising of five generations though it was refurbished with new plumbing, electrical, and windows. Where is the wisdom in forcing out history and us from this stone castle for a freeway where one can race by and then social interaction degenerates into honking and hand gestures (and tweets)." Nick voiced a common sentiment among the foothill communities of individualists living in cobblestone homes.

One would think this a rare virtue. Though many seek stability, the average Californian home owner settles in for less than ten years; then, with historic wanderlust, moves. And for others, the modern motorhome appears preferable to a stable home of stone.

Another beautifully cared for rock home decorated for the Christmas season.

2912 PALOMA STREET

Before the 1930s, in what is now a concrete flood channel, a block east of Nick's home, a seasonal stream flowed in a "large wash". This stone-lined creek shaded by centuries-old sycamore and oak trees suppled the river rocks for Nick's home and this veneered stone home. A November, 1977 *Pasadena Star News* article mentions that Squire Aldridge built this home in the 1920s. Unlike Nick's home, the mason used small stones, an indication that the stones are veneered on a wood frame supporting the stones; whereas, the walls in Nick's home are self-supporting thick stone construction. Notice the wall thickness at the windows of this Paloma Street home. Where Nick's home had thick stone walls, in this home the walls are thin.

1751 ROSE VILLA

Seven years after Andrew Carnegie financed the construction of the 101-inch Hubbell telescope on top of Mt. Wilson, 6,000 feet above the clear skies of Pasadena, this river rock home was built close to where the astronomer Dr. Hubbell taught at the California Institute of Technology, a world famous scientific institution. Albert Einstein also taught at Caltech in 1932 and 1933.

Wealthy people settled in Caltech's neighborhood and built large craftsman style homes. "Only the rich could afford handmade objects," Tom Wolfe wrote, regarding the Arts and Crafts movement of the early 1900s. And craftsman style homes required money. "Rock faced and rock buildings in Europe were signs of upper class, of well-to-do people. 'chunky walls' shows protection and power and strength like castles."[8]

Built in 1924, this beautifully restored home symbolized the owner's status. Remodeled since then, the interiors present a simplified contemporary taste contrasting with the century-old textured boulders of the exterior. Outside old, inside new.

On the west edge of Pasadena is the famous Arroyo Seco canyon where a few cobblestone homes remain. Formerly a wild canyon with a seasonal river, it is now restrained in a concrete channel; but then a stone-covered and sycamore-shaded dry wash, the source of these colorful granite stones.

This stone home's exterior belies its interior, recently refurbished into a contemporary uncluttered fashion, while details like this rusty ash cleaner access door reminds us of wood burning fireplaces of yore.

In 1946, the Arroyo Seco remained a sylvan setting of sycamores and palm trees. The Las Palmas home would be at the base of the bridge down into the Arroyo and just beyond the sycamore. This photo is from the Water and Power Association website waterandpower.org/museum/earlyviewsofpasadena

The historic Colorado Street 'Suicide Bridge' before the installation of the new 210 freeway bridge and the concrete channeling of the Arroyo Seco River

807 LAS PALMAS IN THE ARROYO SECO CANYON

The 1912 Colorado Street bridge spanning the Arroyo Seco Canyon soared 150 feet above us. "Arroyo Seco" is Spanish for 'dry creek' and this great branch of the seasonal Los Angeles River received that name before Los Angeles was founded"[9] in 1781. Under the bridge, a small stream frequently flowed over a spillway, and covered its concrete surface in slippery moss—a perfect steep slip-n-slide if you were a crazy and fearless thirteen year old. Never at a loss to find new and fun adventures, and being of unsound mind, my friends and I in the 1950s rode our bikes down into the canyon to slide down the moss-covered concrete chute. But to get there we gingerly crept on the top of the concrete channel wall over to the stream, careful not to fall, then eased into the stream of cold mountain water. Off we went, no stopping, lots of either delighted or fearful screaming and splashed into the chilly pond below.

I told this story to Ginger, the owner of the Las Palmas stone cottage, and she said, "Yes, another person mentioned a similar story to me. It sounded fun." We reminisced while sitting in her living room and she spoke of her love for her perfectly restored stone cottage located under the looming arches of the Colorado Street bridge. A solitude-loving weaver of fine wool blankets and apparel using ancient looms, Ginger's refinement blossomed in subtle colors and patterns she wove. These muted colors and textures complemented the home's rough and strong stone textures.

Ginger, spry and intelligent in her early 70s, mused, "How did I get this age?" In the early mornings she said she sits in the chair by the tall French doors and looks out to the oak trees, rock wall, and garden of emerald green grass and flowers and reflects, "How lucky I am to live in this rock house."

And she elaborated, "It is an ideal blend (of course she would use the word 'blend') of my woodsy Wisconsin cabin and our city home." She spoke in a gentle, thoughtful manner while her bichon frise rescue poodle, Shatzy, and

I listened.

Upon entering her home, I became immediately impressed with the perfectly balanced positioning of her art and furniture, all complementing each other. Coordinated colors blended with the gray granite stone walls. I had entered the world of a refined and talented artist.

Four years ago Ginger moved here from Wisconsin to continue her weaving art. While touring her two-story home, I noticed both the silence and the complete lack of clutter. She placed every item, whether art or furniture, in its rightful position, not as a staged home to be sold that says no one lives here, but one that breathes of an artful

The exposed stone walls bring a calming serenity and visual delight to the interior.

eye for balance, texture, color, and proportion. A home lived in and loved.

"A special find," Ginger said. "This mid-1800s stone building performed many uses in its life. From a boathouse to an olive oil press building and storage for the Arroyo olive groves. Later it became a barn and stables for an Arroyo mansion, which was destroyed by the building of the 210 freeway bridge. After WWII it served as a temporary residence for a veteran and his family in 1947." The stone structure sat vacant for a while. Then in 1973, the abandoned and dusty stone barn, with its sway-back roof, received its first restoration. A young architect designed the renovation. I was that architect who transformed that empty barn into a two-bedroom, two-bath home for the young owner's residence. The old barn doors were switched to full height French doors, a south-facing dormer added light for the new second floor bedrooms and installing a kitchen and bathrooms completed the home.

In 2004 it was remodeled again and became a high-end residence for a couple who went bust in the 2008 recession. After that unfortunate time, the stone home became a church rectory and office; now it is the private residence for Ginger, a retired weaver who plies her skills on a rare Japanese loom overlooking her manicured garden.

In L.A. city you looked out of your home to see other buildings, streets, cars; in the Arroyo Seco you looked into sycamore groves, streams, and a boulder-strewn wash. In the early 1920s, these canyon residents became known as the 'Arroyo culture' for "they loved crafts such as book binding, writing, wood working, leather work, pottery and metal work...They surrounded themselves with beautiful hand made things."[10] Now a hundred years later Ginger continues this tradition with her artistic wool weavings.

Her aesthetic talent and sensitivity gave renewed life to this unique hundred-and-fifty-year-old stone building. There is something universal about canyons that have beckoned artists and the artistic. We also discovered artists at the south end of the Arroyo Seco where the indefatigable writer Charles Lummis and his neighbors the Browne family of printers and song writers lived in their stone homes. Leaving Ginger's home was leaving a special world, seldom experienced but forever remembered.

There are never too many photos of this sweet and serene rock cottage. A special refuge with many quiet areas.

Upon opening the front door city life returned; traffic noise whooshed by on the freeway high above. Yet inside, quiet, solemn, serenity filled her home. The Vatican library could not be quieter nor more peaceful.

When Ginger purchased her home in 2016 she hired a picture hanger to help out. "I've worked in mansions and very expensive homes," he told her, "But, I've never worked in so special a home like this." And as someone who has designed many homes, I agree. 807 Las Palmas is one special home with a very special owner.

END NOTES

1 Reed., H., *History of Pasadena*, Pasadena Historical Co., 1895, p. 662.
2 Woolsey, R., *Migrants West: Toward the Southern California Frontier*, Grizzly Bear Publishing, 1996, p. 29.
3 Cameron, U.B., *The History of San Gabriel Valley*, dissertation, University of Southern California, 1938, p. 43.
4 Ibid., p. 54.
5 Kipen, D., *Dear Los Angeles: The City in Diaries and Letters, 1542 to 2018*, Modern Library, 2018, p. 66.
6 Sanborn, K., *A Truthful Woman in Southern California*, gutenberg.org, 1906.
7 Reed., H., *History of Pasadena*, Pasadena Historical Co., 1895, p. 387.
8 Wolfe, T., *From Bauhaus to Our House*, MacMillan, 1981, p. 26.
9 Reed., H., *History of Pasadena*, Pasadena Historical Co., 1895, p. 387.
10 Starr, K., *Inventing the Dream*, Oxford University Press, 1985, p. 107.

CHAPTER 9

Forgotten Gates, Watering Troughs, and Waiting Stations

Craftsmanship—"Made for the people and by the people as a happiness to the maker and the user."
William Morris

This three-foot-thick gothic arch is the gateway to a residence in Mountain Home Village, north of Montrose.

WATERING TROUGH, SOUTH PASADENA

Built in 1906, the railroad once carried travelers and freight from Los Angeles to Pasadena and now carries commuters from Pasadena to Los Angeles, and throughout those hundred years it stopped a block north of this waiting station at Mission Street. This two-lane road connects to the nearby San Gabriel mission, built in 1771, and remains the main street of old town South Pasadena. Retired people sit in the shade of green canvas awnings, now, and watch young parents walk their children

The Raymond Rest Station around 1910 on dirt roads. Notice the lady's attire with her flowered hat and the formally dressed gentleman in the shade of the station.

Above: ***Raymond's Stage.*** Above Right: ***Pasadena area as seen in the early 1900s settling on the alluvial plains below the San Gabriel Mountains***

Meridian Street. The South Pasadena Historical Society watches over this way station from across the street in a historic wood two-story building in the style of the Old West with false storefront that extends above the pitched roof. For a few blocks in all directions the pedestrian focus has thoughtfully been retained with small plazas by the train stop, bicycle racks,

a river rock rest station was built for these travelers and the horses. It now rests in timeless peacefulness, shaded by century-old California Oak trees, in the meridian of

sidewalk cafes, restored century-old brick buildings, slow traffic, all shaded and enriched with native California Sycamore trees. A soulful place and a relief to our senses

Above: ***A 1970s photo of the South Pasadena watering station before restoration. The original horse drinking bowl on the left was filled with water; now concrete fills the bowl as all the horses have left. Since this photo, the roof has been repaired with red mission tiles in keeping with the original.*** Above Right: ***The watering station today.***

from the frenzy and ugliness of strip commercial centers on four-lane highways.

A 1988 *Los Angeles Times* article informs us that the "Construction on the trough was begun in 1906, and it was presented to the city on May 20, 1907, to impress visitors arriving at the railway station across the street."[1] It provided a watering trough for tired animals that were used for transportation back and forth to Pasadena during the horse-and-buggy era. The trough was designed by the well-known architect Norman Marsh using the Craftsman style. Its use of redwood beams and large boulders appear to be an exact match to the Greene

The same intersection in 2018, the watering trough is on Meridian Street to the right and the train stop plaza is seen on the left with the table and awning.

and Greene designed waiting station shown next. On completion, the Women's Improvement Association, who funded its construction, placed a bronze plaque above the drinking fountain. It reads: "Watering trough erected in 1906 by Woman's Improvement Association as a rest stop for horses and men as they traveled between Los Angeles and Pasadena."

"The land was leased from the Santa Fe Railway for $1 a year 1906 and the total construction cost came to $364.63," an *Los Angeles Times* article says, and "More than sixty years later the rest stop was restored by the Women's Club of South Pasadena, and recently the tile roof was put back by local Junior Chamber of Commerce members and other civic-minded businessmen."[2]

Though the water trough for the long-vanished horses is now filled with concrete, the artful structure remains as strong as when built. It stands proudly in the original position unhurried, yet unappreciated by drivers and pedestrians focused on going to their destinations.

Scott and I were in no hurry and enjoyed a leisurely lunch at the corner cafe, then walked across the street to the rest station, an old proud centurion refusing to leave his post.

Taken in 1975, this photo shows the waiting station design open to the small park lined with palm trees. Since then, that area changed to a city storage yard for trash and trucks.

OAK LAWN WAITING STATION, FAIR OAKS STREET, SOUTH PASADENA

A century ago the tourists' destination was Pasadena, a new wealthy city of East Coast millionaires and their winter estates. Tourists and visitors arrived by train in Los Angeles and the trolley known as the 'Red Car' would bring the travelers to Pasadena and the Oak Lawn Waiting Station. Located at the border of South Pasadena and Pasadena at the edge of the Arroyo Seco, this landmark was like a bus stop for visitors staying at the Green and Raymond Hotels. Today it has been overshadowed by four-lane highway, a freeway on-ramp, and a large shopping center. And as if to say, "you are not an important piece of history," the city placed a utility yard behind this river rock masterpiece, where storage containers, trash area, and city trucks distract from the rest stop's beauty. It felt a bit like hanging a Michelangelo painting next to a trash can. This once tree-shaded calm respite for local residents, pedestrians, and travelers became a forgotten and invaded relic. A park would have been the wiser and more sensitive choice.

Yet sanity and historic appreciation did squeeze through the commercialization of this area, for in 1988 this modest but bold historic waiting station (South Pasadena Cultural Landmark #3), was catalogued by Arizona State University into the Library of Congress (survey #2290) mentioning, "The Oaklawn Avenue Waiting Station, designed by Charles and Henry Greene and built in 1906 is an amazing combination of redwood timber, brick, cobblestone, and terra cotta tile roof. Located at the edge of the Oaklawn Avenue bridge (also designed by the Greene brothers and now blocked off to cars) and Fair Oaks Avenue, the station was originally built for the passengers awaiting the 'big red cars' on the Pacific Electric Line—a branch of what was once one of the greatest rapid transit systems in the country. The station also partially signified the entrance and set the tone for the handsome houses along Oaklawn Avenue. The 'red car' tracks on Fair Oaks Avenue have long since been removed yet the station stands."[3]

This waiting station's listing in the U.S. National Register of Historic Places in 1973 may have prohibited further widening of Fair Oaks Avenue and the possible destruction of this boulder structure itself.

Raymond Hill Waiting Station, 1900s. A tourist staying at the Raymond Hotel would wait here for the carriage to Los Angeles. Pasadena was a Mecca for tourists during the winter months. The original 'snow birds' would come not from Canada but from the Midwest and Eastern states.

The South Pasadena Watering Station in 2018. Though the water trough for the long-vanished horses is now filled with concrete, the artful structure remains as strong as when built in 1907.

Oaklawn Portal today. Built to impress pedestrians a hundred years ago, today people driving by seldom notice its refined design and craftsmanship.

OAKLAWN PORTAL, SOUTH PASADENA LANDMARK #9

In 1904, a new residential development for millionaires took shape in a peaceful orange grove at the northern side of South Pasadena. Although the tract of expensive and large lots contained only one large oak tree, it was called Oaklawn. "The good life," South Pasadena Realty and Investment Company promised, "is in South Pasadena."

Billed as a "Suburb de Luxe," the posh setting boasted a 75-foot wide tree-lined street, Oaklawn Avenue, with views of nearby sycamore groves, foothills, and rolling hillsides covered with native grasses.

While walking the sidewalks of this wide street, you may imagine its wealthy residents strolling on a warm summer evening to South Pasadena's world famous Raymond Hotel. If that wasn't enough cachet to lure the most discriminating home buyer, Oaklawn's layout and development was designed by the most famous architectural boy wonders of the age: Henry and Charles Greene. Advertisements for the neighborhood described it as a destination "for those who want the best in every particular." Other ads informed potential buyers that "anyone wanting low-priced lots need not apply."

Greene and Greene immediately got to work creating entrance portals grand enough to welcome the subdivision's discriminating new settlers. The resulting structures remain iconic examples of the Greenes' artistry. The clinker-brick platforms give way to river rock pillars topped with large timbers whose edges were rounded to soften their size, handmade roof tiles, and artisan-crafted wrought iron gates. Oaklawn was destined to make a bold design statement, and that statement was right out of the pages of the early twentieth century's flourishing new Arts and Crafts workbook.[4]

THE COBB ESTATE STONE ENTRANCE GATES ALTADENA

Where Lake Avenue refuses to climb the steep slopes of the San Gabriel Mountains any further, two rock gate pillars built in 1918 stand guard to a demolished

The 17-acre Cobb Estate became part of the Angeles National Forest and is open to the public. Photos by Kansas Sabastian

estate. Though not river rock buildings, the gates are included for their unique story and mystical history.

A thorough write up on the paranormalexp.com website mentions that the lumber magnate Charles H. Cobb and his wife Carrie had a large, Spanish-style mansion built for them in 1918. He died in 1939 and the vacant home was demolished in 1959. For fun explore the site in the evening to experience claims of "night-time paranormal activity make(ing) it a popular hot-spot for the curious and those looking for a thrill."[5] This engaging website continues with an intriguing story, "There's been unofficial reports of unexplained screams, laughter, and strange lights." This sounds like a perfect spot to visit on Halloween.

STONE DRINKING FOUNTAIN AT SUNLAND PARK, 8651 FOOTHILL BOULEVARD

In Sunland Park, at Foothill Boulevard and Sherman Grove Avenue, stands this cobblestone drinking fountain built in 1912. Again, we notice the strong influence of women's groups during this time for the 'Ladies Improvement Club" built this fountain for the City of Sunland.

PETER STRAUSS RANCH, LAKE ENCHANTO, SANTA MONICA MOUNTAINS RECREATION AREA

Henry Miller liked fast cars, Indianapolis racing cars. He built many for famous drivers. With the financial success of his racing business in Los Angeles, Miller bought 500 acres in the Santa Monica Mountains and built this stone house in 1926. It burned in the 2018 Woolsey fire. But the 'lookout tower' at the entrance and a stone cabin, located up the hill from the house where Miller had his bootlegging parties during Prohibition, remain unaffected by the fire. A legendary story relates that Henry Miller liked his parties in that stone cabin sited out of sight in the hills. He built the lookout tower at the entrance for one of his employees to warn of any police approaching. I suspect

Above: ***Not shy in the indicating the entrance to his ranch, this stone entry tower, once a lookout tower during Prohibition, now denotes the entrance to Strauss Park. In 2018 the Malibu fire burned through this area destroying homes, bridges, and trees. The tower and the rock walls of the home withstood the conflagration.*** Above Left: ***Original rock home of Henry Miller at Lake Enchanto.***

Top Left; ***Historic photo of the entrance to Lake Enchanto.*** Middle and Right; ***The remains of the ranch home after the Malibu fire in 2018.*** Above; ***The native stone amphitheater showing part of the concrete stage on the lower left.***

he bought the property in this very remote location with little police presence primarily for these parties. Ever the racing man, Miller didn't retire from life's fast lane and stayed in the auto racing business and continued the boozing parties.

But the Depression years resulted in financial ruin for him, forcing the ranch's sale to its second owners who improved the ranch over the 1940s into a daytime fun zone and resort named Lake Enchanto. Families and children frolicked, picnicked, swam, and hiked to their delight for a couple decades at this local attraction.

An enormous swimming pool was fed by the Triunfo Creek, which meandered through the ranch. The creek continued down the ranch to form a lake where children enjoyed fishing and tossing pebbles. Outdoor picnic areas and trails were built for children to explore, with their parents in tow, and an amphitheater where legends like Johnny Cash and Willie Nelson played during outdoor summer concerts.

In the 1960s it fell in popularity, and Peter Strauss, the award-winning actor of *Rich Man, Poor Man*, stepped in to purchase the ranch. His vision returned the property to its more native condition, without demolishing any of the stone built improvements. A few years later in 1983 Strauss moved to Ojai with a new gardening project. He sold Lake Enchanto to the Santa Monica Mountains

Right: ***The Ranch Land in the Santa Monica Mountains***

"Early day drinking fountain on Olive Street," the handwritten note tells us.

General merchandise mixed with produce and stables in the early twentieth century grocery store. Named the 'mercantile stores' their size was quite smaller and more compact than today's grocery stores with aisles for large shopping carts .

Conservancy, and they improved and maintained the ranch in excellent condition as seen in the before the fire photo of the stone house.

For ninety-two years, 1926-2018, Lake Enchanto flourished under the care of three unique men of vision. During those decades it survived a few earthquakes, a few fires, and with the resulting winter floods, but the Woolsey fire finally got to this beautiful stone home. And noticing the wood supporting the stone walls after the fire, I hope that it can be rebuilt for future enjoyment and history.

Unexpected old photographs have shown up occasionally during our research, and this one of a now-demolished stone drinking fountain for horses illustrates the wealth of many of Southern California settlers who built their homes in the San Gabriel Mountain foothills.

The handwritten notes describe this as an "everyday drinking fountain on Olive Street." That would place this in Glendale. We need to fast rewind to the late 1800s when horses pulled wagons and carriages. Why the tall tower part? Was it a cistern for the water, was it tall so a rider and carriage driver could see it from afar? People the practical nature would not have spent their days to build it tall without a purpose. The background shows a wood-framed mansion with steep gabled roofs, as if Glendale was snow country. This indicates that the home's owner probably immigrated from an eastern U. S. city of snow and industry to build a mansion for his retirement.

END NOTES

1 De Wolfe, E., "A Watering Trough Remains," *Los Angeles Times*, February 21, 1988.
2 Ibid.
3 Library of Congress, www.loc.gov.
4 Allee, L., "Exploring Oaklawn," *South Pasadena Patch*, January 18, 2011.
5 Paranormal Experiences. www.paranormalexp.com, accessed on January 12, 2018

CHAPTER 10

Sierra Madre, Santa Anita Canyon, and Al Foulds

1873: As I raise my eyes...and gaze toward the north, how the brow of the Sierra Madre glimmers and glitters in its robe of snow!...Here winter and summer stand face to face....The future wealth of this country will be its tropical fruits. An orange, lemon, lime, or walnut orchard, is better than a gold mine.[1]

Edith Bowen wrote of her interview with the aged Mrs. Carter, the founding mother of Sierra Madre: "She looked out of the window to the south, over the wide sweep of the San Gabriel valley. 'It's not the same, even in daytime,' Mrs. Carter said. 'Why, I remember when we could see the Pacific every day. It was like a silver ribbon way of there. But the cities have brought smoke to screen it and we don't see it as often now...We came from Lowell, Massachusetts...my husband was ill and he was advised to come West. They called it consumption in those days.' Mrs. Carter continued her story of when they arrived in Southern California in the 1880s. 'He weighed 134 pounds when he came here, and when he got well he weighed two hundred pounds. He was so pleased that he went back East to show them. He talked so much about the climate that many families from New England came here.'"[2]

Source; Sierra Madre Library and Historical Society.

Even in this sublime setting practical issues needed to be dealt with, and a fledging community far from a big city had some unique problems. Again Edith Bowen's work shares a few: "ordinances were passed at every meeting," Mr. C. W. Jones writes in his letter, 'and many of them were of a trying nature, such as the one eliminating tent houses, controlling burros on the public streets, regulating sanitariums and their practical prohibition, within specified zones.'"[3] Today many a politician and citizen may wish they had such issues of a 'trying nature'.

Mrs. Carter died during her ninety-first year, and earlier she and her husband survived the tuberculosis pandemic of 1917–18 when "People are talking about

In the early 1900s, this shows the sparse settlements on the San Gabriel foothills.

the inconveniences of theaters and shops being closed, or people wearing masks and how they will look. Then it becomes commonplace because everyone is doing it."[4] These feelings are to be respected in order to understand the culture of the early 1900s when a few river rock homes were built.

Mrs. Francis Hawks, another of Sierra Madre's early settlers, tells us, "The life so simple and peaceful, though not exactly restful at all times...we would stand and admire the mountains, which mother said seemed as a sheltering arm protecting us from all harm...the early sunrises were pictures of great beauty. Each day brought new delights as well as added cares but my health was returning, which we considered the greatest recompense for our labor and self-denial. There's but one Sierra Madre - that has no rival."[5]

Earlier we asked, "Why use stone for the homes and buildings?" Mrs. Hawks may have one reliable reason as she mentioned in a 1950 book on early Sierra Madre: "We soon found workmen to batten down our room (a one room house made with vertical wood planks and joints covered by wood battens) and build two more rooms. With this change came a never to be forgotten surprise in the form of a fierce wind storm lasting several days and nights. A new chicken house had been built and posts set firmly in the ground, and there we spent our night, feeling it safer than our house."[6] Stone homes were not to be blown away, like many a house, tent, or wood-framed building. Stone walls also kept out snakes and dust. Both frequent in the founding days before being reduced by agriculture and asphalt.

In the late 1800s, Mrs. E. T. Price shared her feelings on being a settler on new lands, a feeling that is impossible to experience in these lands of freeways, traffic, strip malls, and millions of people.

"'In their place (city development and roads) you must see a land of almost primitive, pastoral simplicity... We were young and enthusiastic and animated by a spirit of confidence and hope and high endeavor, our hearts turned lingeringly toward the friends and loved ones we had left 'back East.'"[7]

Though growing up in Sierra Madre, I was unaware of its rich, but typical, heritage, where, as Edith Blumer Bowen wrote,"Sierra Madre rose on the decomposed granite outwash plain below" the San Gabriel Mountains.[8] Sloping land "mostly carpeted with a pungent chaparral mix: sagebrush, sumac, poison oak, and native grasses."[9]

N. C. Carter's story was similar to many other men of means who settled this new land. "With his health quickly and dramatically improving (having recovered when living in the climate of the foothills) he was able to return east in the following spring, in 1872, to bring" his wife and children by wagon from Massachusetts out to Southern California....Carter...plunged into his new career: real estate...beginning in 1873 or 1874, he went back east to promote westward immigration and land sales. His promotional literature included before and after photos of his healing all due to 'The Climate of California' to support his effort in bringing friends out to Sierra Madre."[10]

In 1869, the trans-continental railroad was completed, and Mr. Carter could then advertise a "pioneer excursion party" of 40 to 60 people to ride the new train from Boston to Los Angeles for $50. A few years later the Southern Pacific and Union Pacific would enter a fare war lowering that price to $1 in 1880s. Though Mr. Carter was excited about sharing his renewed health, selling the one thousand acres he bought may have been a strong motive for his promotional efforts.

"When the Sierra Madre Tract was purchased, within that tract lied the Little Sant Anita Canyon Wash, an asset that attracted Los Angeles builders. Said asset being the huge granite boulders that had been carried down the Canyon through the ages by cloudbursts and heavy freshets (flooding rivers). These proved to be valuable for retaining walls, foundations and building of office and store buildings."[11] This entry by Stella Dennison remains one of the rare statements about the presence and value of the river rocks for building.

This early plein air (outdoors) painting depicted the foothills of the San Gabriels before people, roads, buildings, and cars. The oil painting by Armand Cabrera in the book by Michele Zack, **Southern California Story: Seeking the Better Life in Sierra Madre.**

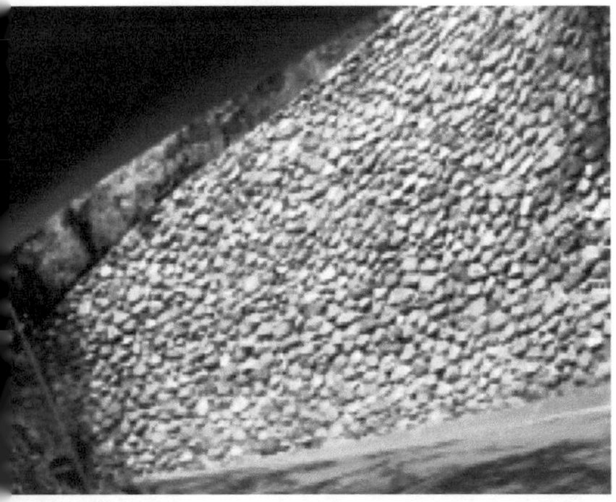

ASCENSION CHURCH

In the late 1800s an Episcopal church in Sierra Madre was built with wood. It took "Four months to build and furnish and install the stained glass windows," Mrs J. Hart mentions during a midcentury interview.[12] "A short time," she said. But there were no city inspectors or building permits to deal with. And you might understand or not be surprised that it blew down having been built in such a brief few months." The strong Santa Ana winds destroyed this new church.

"All was going well," she continued, "until the night of October 10, 1887, when an unusually heavy wind came. The sides of the church gave way, and the building collapsed...A meeting was held on the ruins the next afternoon. It was decided a larger and more substantial building should be erected on the same site...At a cost of five thousand dollars, a stone church, designed by Ernest Coxhead was erected. The stones for the building were hauled up from the Little Santa Anita Wash by voluntary labor."[13] Though a mere six months was needed to complete the stone church, two months longer than the first wooden one, it has stood for over one hundred and thirty years, not four years. River stones became the preferred material for churches built in the foothills.

In the Sierra Madre canyon roads river rock constructions appear at every turn, not only homes but large rock retaining walls along the narrow winding roadways with mature trees shielding you from the hot

15-foot-high stone retaining wall at Oakdale and Sturtevant Drives in Sierra Madre canyon. There are no cracks though it was probably built in the 1920s or earlier.

summer sun. Here there are no sidewalks and definitely no parking on the sides of the street. Timid drivers crawl along at 10-15 mph and a more aggressive speeder may venture to 25 mph. But you are there to marvel on the quiet scenery of boulders, trees, river rock buildings, and large rock retaining walls, just a few minutes distant from city turmoil.

Bethany Temple in Sierra Madre, photo from historic postcard illustrating its unique beauty.

BETHANY TEMPLE

In October, 1922 *Popular Science Monthly* published a half-page article on the recently completed Bethany Temple. Hidden among articles describing recent mechanical inventions, the magazine highlighted this building's unique and, in many ways strange, invention created by one man. All the sections of the building were circular and constructed of river rocks from "near-by streams". The article mentioned that this semi-blind stone mason, Mr. Cornuelle (1864-1945), built Bethany Temple in one and a half years—alone. He had settled in Sierra Madre after immigrating from France and was known locally.

The dome in the congregants' wing soared above them spanning around sixty feet and supported only by the circular perimeter walls. Domes are a self-supporting shape, as seen in many European cathedrals hundreds of years old. The magazine's complete article, shown on the next page, describes the intriguing methods Cornuelle used in building the Temple. Clearly he was a master of his trade.

Its name, "Bethany Temple" is written below the high arch above the door.

But the distinctive round stone temple was not unique enough to prevent its sad ending. The rocks and

One Man Builds Domed Church

COMBINING in its wide sweeping dome, its windows, and cornice decorations, interesting features of both Greek and Norman architecture a unique religious edifice under the name of Bethany Temple, has been constructed in the city of Sierra Madre, Calif., by one man. Nothing about the edifice is professional. All the materials were taken from near-by sources.

The temple proper is 52 feet in diameter and 30 feet high from ground to top of dome. A second building, which houses the Sunday school, is 57 feet in diameter and 18 feet high. Both structures are built of cement blocks with a facing of smooth granite stones lifted from a near-by mountain stream.

Extensive as the buildings are, the entire construction work was done by a local artisan, L. D. Cornuelle, under supervision of the Rev. W. H. Rawlings. It required a year and a half to complete the structures.

The sweeping arches of the church and the semicircular windows under the eaves are shown in the interior view at the right. Note the excellent finish of the auditorium

Gothic windows and graceful domes with walls of stones taken from near-by streams were selected by the designer to distinguish Bethany Temple, a religious center at Sierra Madre, Calif.

At the right is L. D. Cornuelle, who constructed the buildings. His ingenuity enabled him to erect the walls and domes alone and without the customary extensive staging

In building the temple and school, Mr. Cornuelle adopted an unusual method of scaffolding. After the course had been carried to a point as high as he could reach from the ground, a spring wagon was drawn alongside and used as the staging. When the new height had been overreached, a platform was erected and up this the wagon was drawn. As the tier spread to one side or the other, the wagon was shifted along the platform. Later a staging was built upon the wagon and this gave the necessary height for the completion of the walls.

Only the dome is of wood; all else is concrete. Departing from the usual method of dome construction wherein the rafters supporting the dome are sawed to shape, Mr. Cornuelle bent one-by-three timbers to the required curvature and then nailed several of them together to give rigidity. Over the rafters he placed lead covered asbestos paper well sanded with broken gravel to match the concrete and stone of the walls.

The edifice is illuminated by indirect lighting in invisible fixtures in white and soft colors that can be manipulated at will.

domes of Sierra Madre's Bethany Temple were demolished a few years ago, and on the site a very bland, uninspiring, mammoth church was built, looking like an elephant in a duck pond, dwarfing Sierra Madre's small-town ambiance.

Edith Blumer Bowen wrote in her book that "Father Barth began Catholic services about 1907. Bethany church and the Jewish congregation both appeared much later."[14] The main auditorium of Bethany Temple, which I'd judge to be the circular building behind the main sanctuary, is said to have been fifty-two feet in diameter and thirty feet high, making this one of the most imposing of stone buildings.

It's a head shaker trying to understand the insensitivity that allowed the temple's destruction. This magnificent and one-of-a-kind stone temple with its rich history demanded to be preserved. Sierra Madre and its citizens suffer for it demolition.

SANTA ANITA CANYON

Although it is only 20 minutes from downtown Los Angeles, it seems far removed not only in space but also in time.[15]

A boy in red pants floats on a log in the canyon pool. Seventy-five-foot monolith of gray granite Sturtevant Falls fills this cool pool. This is not a boulder. Boulders have edges, but this sheer-sided granite outcropping forms a small part of a granite mountain that is thirty-five miles long and fifteen miles wide, 10,000 feet high, and still growing. I sit next to the pool diminished by the monolith's height and unfathomable power. This celestial canyon rivals Notre Dame Cathedral in wonder, sound, size, and beauty, an experience of awe and humility.

Here the mountain's sides embrace you with nature's peaceful, calm hug. Yet here lies a small piece of of nineteenth

Boy straddling a log in the pool below Sturdevant Falls, Santa Anita Canyon.

At the bottom of the canyon this narrow and uneven trail is the only route to the cabins, as it has been for a century. Imagine walking home in this idyllic setting of shaded coolness.

century Southern California hidden.

Three hundred fifty stone cabins were built in this canyon from 1909 to 1919, and eighty-one remain in this off-the-grid canyon village. Floods in 1934, 1938, and 1969 destroyed many, and then mountain fires in 1953 removed a few more from this historic place of refuge, a place where the trees rustle in the breezes, the stream gurgles in its laziness, hikers soothe their feet in the cool mountain pool while their calming their souls in this bucolic canyon.

A lengthy hike down the mountain side leads to the canyon bottom where the cool forest of sycamore, alder, and oak trees live. A tree-shaded trail follows the stream to the village of one-hundred-year-old boulder and river rock summer cabins.

Food and supplies? There are no cars here, just mules. From the pack station in Chantry Flats, at the top of the hillside, mules were loaded with supplies and furniture then led down a steep and long trail, cross a bridge or two, depending if the wooden foot bridge survived the heavy storms that year, to deliver the goods and groceries to your and your neighbor's cabins. Home delivery service here, then and now, and not much has changed in one hundred years when people relied on their neighbors. Community was survival.

Al Foulds built most of these stone cabins while in his mid-forties, and in 1968, while living in Sierra Madre he celebrated his ninety-fifth birthday. He did not celebrate his ninety-sixth.[16]

Foulds describes building the cabins in 1912 through 1916 in a 1968 letter to his grandson's wife. "I went into the Big Santa Anita Canyon in the summer of 1914 for a vacation as work was not plentiful in L.A. I was a stonemason by trade and there was lots of it to do in the canyon. I stayed there; I built what was known as First Water Camp for a man. In time business fell off, he walked

Al Foulds, petting his friend's dog Betty, wears a dark shirt with leather suspenders and an easy smile. He sits with friends Otis Johnson and Art Mills next to his rock cabin posing for a 1930ish photo.

out...By that time I became one of the all-time occupants of the canyon."

Foulds bought the cabin "for what his client had put into it and lived there...right across the crick (sic) of the ranger station."

He continued, "The canyon was opened up for cabin sites in 1912, so it was quite new. (The Forest Service rented the sites for summer vacation stays, with a few full-timers living there.) Then there (were) some people, mostly young people, climbing around the cliffs (who) would fall and hurt themselves that bad they had to be carried out by stretcher. It took all the men that were there to do that... Then someone would die and had to be carried out to Sierra Madre. There was no road then, just a horse trail; everything (that) came in or out was done on a pack train. But (these things) only occurred once in a while."[17]

Communication then and now occurs one of two ways: either hike up the steep trail to Chantry Flats or use the original hand crank telephone, the only phone in the canyon.

During our visit I felt a time warp, so close to cities yet so remote in time from the cities. Being down here, stopping and listening to the birds, the breeze rustling the sycamore leaves, and the creek's gentle gurgling carry you back to Al Fould's time.

With trekking poles, a bottle of water, and camera, we hiked down that steep old mule trail, now a paved path, resting at a few spots shaded by the centuries-old canyon oaks. Hikers puffed their way up stopping

The steep trail into the Santa Anita Canyon and to the river rock cabins. Original summertime settlers leased land from the National Park to build their cabins in the cool canyon. Stout hikers these settlers must have been to traverse this trail, lead a pack train of donkeys to their cabin with building and living supplies, and then hike back up to the trail head.

Left: ***Cabin 58, one of the few two-story cabins and where we met Trevor its owner and full-time occupant.*** Top: ***The kitchen/eating room on Trevor's first floor. This river rock cabin has a second floor, very unusual for these rock cabins.*** Above: ***You can imagine sleeping in the nighttime canyon open to its cool breezes filtering through the screened second floor.***

to chat about the canyon's beauty or ask for directions, for other trails branch off and, for strangers, it was easy to get lost.

A half-hour later we reached the canyon floor. Walking among the cabins we met one survivor, cabin 58, and its new owner, Trevor, a robust young man sharing his dream with rattlesnakes, mice, and day hikers who wander by his canyon cabin. In this bucolic setting, neither for the fragile nor the faint-hearted, night-time silence pierces your soul.

Trevor was repairing his new home, the only two-story river rock cabin in the canyon. And in the fashion of Al Foulds, Trevor was joined by Jake, a 20-pound terrier who helped by instantly befriending strangers and wanting nothing more than a gentle ear scratch.

A few years ago Trevor arrived here to help an older man restore this cabin. The man, before his death had asked a canyon friend, his neighbor, who he should sell the cabin to. Without hesitation the neighbor replied, "To

Top Far Left: ***Trevor and I chat while looking out the second floor screened openings.*** Top Middle: ***The outhouse. Can you imagine using this at night?*** Top Right and Above: ***Typical century-old stone cabins, still in use. A small piece of the early twentieth century hidden in Santa Anita canyon above Sierra Madre.***

Trevor, of course. Why he's helped the past years and is dedicated to keeping the cabin goin'."

These cabins are not listed on realtor sites since they are passed down generation to generation within the family. Trevor received a rare find, for the man's estate allowed him to pay a reasonable mortgage over time.

MARY'S MARKET 561 WOODLAWN DRIVE IN SIERRA MADRE CANYON

Imagine being awakened and frightened by the deep rumbling of boulders and rocks tumbling down your street, rain drumming on your roof, your modest wood canyon home shaking in the violent winds. A resident during a torrential storm of the early twentieth century had cause to wonder if his world was closing in, if he was safe or would be swept down the canyon. Such storms occurred once or twice a decade, yet the canyon people persevered and survived.

This appears to be an early 1920s photo based on the cars. The Spartan grocery store is now Mary's Market, a sublime setting for this rural canyon cafe.

And so did Spartan grocery store which provided food and basic supplies for these remote canyon people. Through the years the name changed to "Mary's Market". Today it's a small hidden sandwich shop, revered by the locals for its location among the many canyon homes that were built decades ago for the 'invalids' (people with tuberculous) seeking dry air and nature's cures.

This area has remained unchanged over the decades and became part of our journey into the area's past. After photographing the river rock cabins in Santa Anita Canyon, we ate lunch here, rested under the giant sycamore whose broad leaves rustled in the breeze. It was impossible to be stressed or anxious. There we planned our afternoon trip to a singular stone castle.

River rock buildings were not confined to the canyons or washes, an outlier in time, location, and size sits a few mile east of here built by a man who had trouble

Placed on top of this low rock wall a century ago, these mailboxes still serve the hillside houses in Sierra Madre canyon. Sitting outside Mary's Market, we watched as a six foot mailman, carrying his leather mail bag, bent over to place mail in each box. He didn't complain; he didn't report this low level of the boxes to his supervisors. He is part of this canyon's historic culture and its rhythms—unchanged for decades—a pre-electronic sylvan life.

solving math problems at school, but had no problem building a stone fort.

END NOTES

1 Kipen, D., *Dear Los Angeles: The City in Diaries and Letters, 1542 to 2018*, Modern Library, 2018, p. 445.

2 Bowen, E.B., *Annals of Early Sierra Madre*, Sierra Madre Historical Society, 1950, p. 41-42.

3 Ibid., p. 44.

4 Klein, J., "A Trove of Sad, Funny, and Familiar Stories from the 1918 Flu Pandemic," *Atlas Obscura*, May 2020.

5 Bowen, E.B., *Annals of Early Sierra Madre*, Sierra Madre Historical Society, 1950, p. 49.

6 Ibid., p. 48.

7 Ibid., p. 71.

8 Ibid., p. 81.

9 Ibid., p. 81.

10 Ibid., p. 82.

11 Ibid., p. 84.

12 Ibid., p. 115.

13 Ibid., p. 116.

14 Ibid., p. 41-2.

15 "Al Foulds," Big Santa Anita Canyon, www.bigsantaanitacanyon.com/stories/foulds1.htm.

16 "Big Snow of 1922," Big Santa Anita Canyon, www.bigsantaanitacanyon.com.

17 Letter by Alfred Foulds, 1968, Big Santa Anita Canyon, www.bigsantaanitacanyon.com/stories/foulds1.htm.

CHAPTER 11

Glendora, Rubel Castle, and a Little About Building Codes

There are no rules of architecture for a castle in the clouds.

Gilbert K. Chestorton, *The Everlasting Man*

RUBEL FANTASY FORT
844 NORTH LIVE OAK DRIVE, GLENDORA

East of Sierra Madre, east of Monrovia, further east past the towns of Duarte and Azusa, in the San Gabriel Mountain foothills, stands the celebration of a childhood dream.

Michael Rubel, in the latter part of the twentieth century, seeking fun, built a river rock fort reminiscent of medieval castles. Rubel, an outlier, and his friends enjoyed themselves in the hard work of creating this castle, which has become a historic landmark. Rubel did not have a huge inheritance, but his riches lay in having a lot of fun-loving friends and an influential mom.

Rubel was an uncommon man in a quiet foothill town, befriending all with his natural, wide smile set under dark rimmed glasses that slipped down his nose, giving kindness to his friendly face. I can imagine the full, dark brown mustache dancing in rhythm with the jiggle of his double chin and full body when he laughed, which probably was frequent. Michael had friends, some say a thousand or more, and they would join him on weekends to escape their work-a-day lives to help build this fort.

Top: ***Entrance drive at 844 N. Live Oak, Glendora, Rubel Castle.*** Right: ***The fort's main gate lowers to close the entrance. Every fort needs a strong fortified gated entrance.***

Above: ***Michael Rubel*** Right: ***Cobblestone clock tower with the clock and bell from England***

Starting in 1960 and for more than twenty-five years they hauled stones and timbers and mortar, all seeing their stone creation as fun.[1] At twilight as darkness approached and work ceased, they cleaned their dusty, chapped hands, looked proudly at their work that day, and relaxed with a barbecue, a few beers, and many stories of the day's efforts.

To Rubel and his friends this was play, a magical day building their dream, a fort complete with fake cannons of short blackened telephone poles high in the parapets of rounded stones, a drawbridge protected by sturdy circular stone walls, and an authentic British bell tower that tells the town the hour to this day. "In the end," Glendora Historical Society Museum's pamphlet says, "he wound up following his boyhood dream and compulsion of building a castle. So, with river rocks, abandoned cable, railroad track, and his wide circle of friends, Michael began building the castle in the old reservoir. It took... years of hard work, perseverance, and frivolity to create Rubelia."[2]

Rubel was an average student, but his mom was well known and successful in Glendora's commercial real estate, an important component to his story, as we

These steps are strong, but a handrail would be nice. An inside view of the fort with the drawbridge on the right of the stairway. The castle was not designed on plans but grew out of of Michael Rubel's fantastic mind. Touring the structures built inside the reservoir

learned when we toured the castle. Towards the end of our tour (call the historical society to make reservations), we watched a short movie on Rubel's life and work and learned of his self-effacing hubris: "I couldn't carry grades in school very well, but these stones...they weren't too heavy." It was 1959 when he bought an old citrus tract of 2.5 acres, but he had no intention of farming. He wanted the concrete reservoir. It provided the skeleton of his castle onto which he grouted the river rocks and then built many separate stone structures within—one like the clock tower which rose to 48 feet. Michael Rubel died in 2007 leaving the lasting heritage of his fort to the Glendora Historical Society and to our delight.

The inner court with tower.

Rubel's relics.

BUILDING CODES AND EARTHQUAKES: CLAY BRICKS, GRANITE STONES, AND MORTAR

"Always a factor in California, earthquakes are something every Californian must live with but that few care to think about."[3]

We journey back to March 10, 1933, a Friday afternoon, and to another town to continue Michael Rubel's story. Long Beach was a young town in then whose citizens were unaware that an earthquake would jolt and toss them around like children on a trampoline. A 6.4 magnitude left the city in ruins—demolished schools, hospitals, civic buildings, and homes. Most of these demolished buildings were constructed of bricks. They collapsed. Total failure. Brick walls crumbled, roofs fell, cars crushed, and people where injured by the flying bricks. This man-made material breaks with ease and crumbles into sharp-edged pieces. The collapsed buildings were built of red clay bricks stacked on top of each other like boxes of shoes in a shoe store, then glued together with a weak mixture of lime, sand, and cement, a cheap mortar.

Yet a two-story river rock house built in the 1920s at

A brick building destroyed by the 1933 Long Beach earthquake.

Early 1920s stone home on the cliffs of Long Beach with its owner possibly waving good-bye to a friend. The home stands today, undamaged by the earthquakes.

Freeway collapse in 1994 earthquake, the intersection of 14 and 405 in Santa Clarita.

the top of the ocean cliffs in the same city, Long Beach, experienced no damage. "In the early twentieth century structural aspects for building design" a Department of Housing report states, "were not codified or standardized. Houses and their structural members (beams, columns, walls, and foundations) were selected, chosen by heuristics, that is by experience and rules of thumb."[4]

In that Long Beach earthquake, adjacent cities like Inglewood, Lawndale, Torrance, and southern Los Angeles experienced extensive damage to their civic buildings, commercial stores, homes, and streets. The trauma and panic from this massive event captured the immediate attention of state and city engineers, the people responsible for writing building codes. In 1935, only two years later, quite a short time for changes to the code, structural requirements increased and unreinforced—that is without steel reinforcing adding tensile strength—masonry was disallowed.

These building code writers were most likely thinking of bricks, yet river rocks got caught up in the classification of 'masonry'. I think this was unintentional as the codes did not (and still do not) ever mention the words 'river rock' or 'cobblestone'. There was no acknowledgement of that type of construction, maybe no understanding, that the strength of nature's granite river rocks far surpasses the strength of man-made clay bricks. Possibly the structural code writers overlooked or ignored acknowledging the proven structural integrity of cobblestone homes, the homes with 18-inch-to-30-inch-thick self-supporting stone walls. As described in a previous chapter, river rocks bound together with Portland cement mortar results in a monolithic structure and can ride the rolling grounds with ease like a ship in the ocean.

Slam a brick over the edge of another brick and the brick breaks. Slam a bowling ball sized granite river rock over another and only a loud noise results as it ricochets and bounces away. Like an assembled jigsaw puzzle, these river rock buildings became monolithic, held together in one whole with irregular and discontinuous joints. Brick buildings, however, have easily fractured straight lines, the flat mortar joints. Hold an assembled jigsaw puzzle horizontally, you get the idea, it doesn't fall apart, not one piece drops.

During the twentieth century Southern California experienced many sizable earthquakes: on July 21, 1952, a 7.3 seismic event south of Bakersfield bent railroad tracks and frightened most everyone in the L.A. basin; then on February 21, 1971 a magnitude 6.6 event shook Simi Hills demolishing a newly constructed hospital, which had not yet been occupied; October 1,1987 a 5.9 event at Whittier Narrows; in Upland on February 28, 1990 a 5.8 quake occurred; and next year in the early morning of July 28, 1991 in Sierra Madre a 5.6 event; and at 4:31

Home at 1100 East Foothill Boulevard and Pfleuger Avenue.

Cornerstone Bible Church, formerly Brethren Church, Glendora, 400 North Glendora Avenue, Glendora.[7]

on January 17, 1994, the most costly earthquake in the U.S. a 6.7 event at Northridge in San Fernando Valley shook all of Southern California resulting in billions of dollars of damage and a major rewriting of the structural earthquake codes and the insurance policies. These seismic events and a few more, startled Californians, destroyed many buildings, injured and killed hundreds of people, and collapsed bridges that California freeway engineers thought safe. "In just twelve seconds, nearly 70 bridges designed to resist earthquakes...and thought to be quake proof were damaged." In his 2014 book, *Earthquake Storms: The Fascinating History and Volatile Future of the San Andreas Fault*, John Dvorak reminds us that "Seven freeway bridges collapsed."[5]

But none of these river rock buildings failed. A few had slight cracks or loosened stones. An unsupported chimney may have toppled a few stones, while most chimneys remained intact. Buildings of veneered stone, walls three to four inches thick, were not included here, and many of those buildings failed. "In a thin veneer wall...the stones are all set on edge. The result is an appearance distinctly different from older, traditional, solid wall construction."[6]. It could be that the engineers confused or conflated the two structurally different constructions.

Solid stone buildings, held together by Portland cement mortar, were built by stone masons who may have known more of their trade than the engineers who wrote the building codes. Stone masonry was, and is, rooted in the practical, not the theoretical. Their skills left us unique, enduring stone buildings from a hundred years ago. Yet construction standards and tastes have left these stone structures behind, with many ignored. Try to obtain city approvals to build a 16th century stone cathedral, even on a smaller scale. The governmental system is not set up for that. They do not know the technology of a cathedral's design. And on a much smaller and local scale, government agencies do not understand the unique strengths of river rock structures, so getting approvals to build one today is impossible.

Engineers and building officials focus on failures. An important aspect for our safety, but this approach overlooks successes. In earthquake country it would appear that the buildings which survived undamaged would be just as important to learn from as those that failed. What worked combined with what failed yields growth and success.

There was another issue at play. Except for Rubel Castle, most of the stone buildings were built prior to building codes, so it was up to the contractor to provide a strong, lasting structure or he was out of business, either by word of mouth or reputation and he would be avoided. In those small foothill communities, word traveled throughout the towns as swiftly as the blowing dust. And how did Rubel get by these 'masonry' codes enforced twenty-five years before he built his castle?

I mentioned earlier that he built the fort with the help of friends and, importantly, his mother, a successful commercial real estate broker who was well-known in that small Glendora community.

RETURNING TO RUBEL'S STONE CASTLE

A few years after Michael Rubel started his river rock fort, city building inspectors became interested. He was building without a city permit. That upset them.[8] Mom then became involved. And she had friends, many, many friends. So did Michael. And his friends' mothers lived in Glendora. During a City Council hearing on the matter, Mom arrived along with his friends and their mothers. They filled the council chambers and overflowed into the lobby and halls. Mama Rubel spoke to the City Council and the presiding Mayor. She suggested that they "leave Michael alone" should the fine councilmen want to be re-elected.

Rubel and his friends continued building unmolested for twenty more years and his stone 'castle' outlasted all the councilmen and the building inspectors, to become a lasting illustration of how practical knowledge and skill can be smarter than textbook theory and inexperience. Rubel Castle's soaring towers, these unreinforced structures, have remained undamaged, while the nearby Sierra Madre and Northridge earthquakes destroyed many structures built to the prevailing codes.

Details of Rubel Castle, a monumental achievement of one man and his many friends.

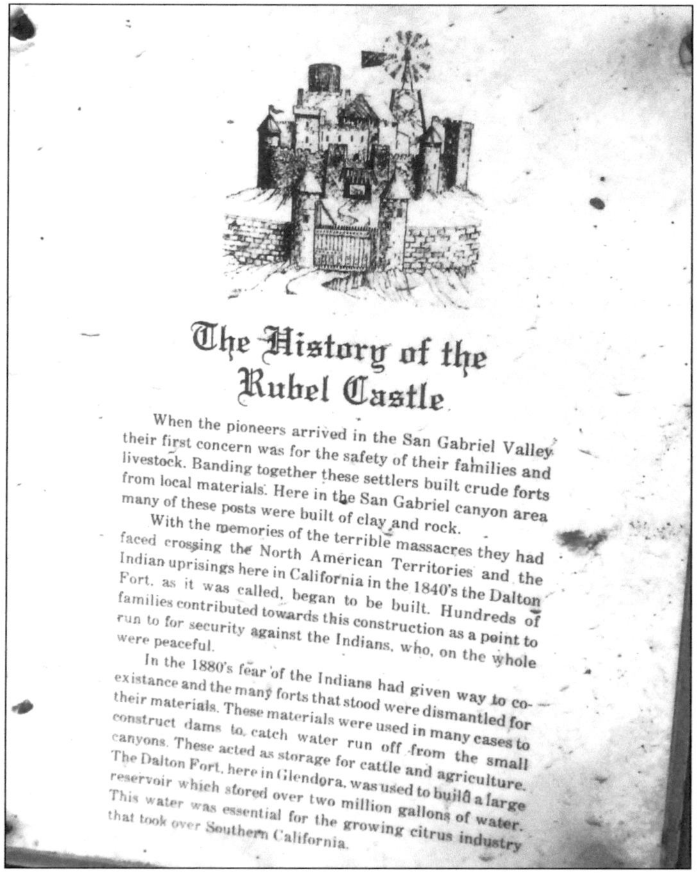

Michael Rubel created a rare marvel for the ages, like the Watts Towers in Los Angeles or the Winchester Mystery House in Santa Clara. Rubel didn't look for fame or posterity. He was interested in having fun; throughout his life he remained a boy at heart.

And though they may be labeled 'eccentrics', the Rod Serlings, the Walt Disneys, the Howard Hugheses, the Mrs. Winchesters, and the Michael Rubels, all were visionaries who realized their dreams and pointed the path to personal fulfillment and happiness. Much can be learned from these great imaginative visionaries.

END NOTES

1 "Rubel Castle...One Man's Dream," pamphlet, Glendora Historical Society Museum, March 17, 2012.

2 Ibid.

3 Hayes, D., *Historical Atlas of California: With Original Maps*, University of California Press, 2007, p. 233.

4 "Review of Structural Materials and Methods for Home Building in the U.S., 1900-2000," U.S. Department of Housing, huduser.gov.

5 Dvork, J., *Earthquake Storms: The Fascinating History and Volatile Future of the San Andreas Fault*, Pegasus Books, 2014, p. 10.

6 Long, C.K., *The Stone Builders Primer*, Camden House Publishing, 1981, p. 13.

7 *Los Angeles Herald*, UCR Center for Bibliographical Studies and Research, March 6, 1910.

8 *The History of Rubel Castle*, movie, viewed in 2017.

CHAPTER 12

City of Claremont and More on the Domestic Immigrants

Claremont has some wonderful stone bungalows built with Claremont potatoes (cobblestones)

PITZER RANCH,
4353 NORTH TOWNE, CLAREMONT

"On the northern slope of Claremont," Elizabeth Pomeroy writes in her guide book on Southern California Landmarks, "the Pitzer House faces the morning sun and eastern peaks (of the San Gabriel mountains) as it has since 1912. This picturesque fieldstone bungalow has been called by architectural historian Robert Winter 'the finest stone house in Southern California.'"[1]

Long branches of the centuries-old California oak tree sheltered and framed his river rock home. Both were expansive and horizontal, as if the home followed the oak tree's form adapting to the SoCal heat. To give relief to its inhabitants from the summer heat, the thoughtful early building design used overhangs, cross ventilation, screened sleeping porches, and low horizontal profiles with thick walls and porches. It remained cool inside.

On January 1, 1901, Lee Pitzer married Blanche, the daughter of James Becket who owned this land. The eleven rooms of this 3,000-square-foot home surround a central courtyard providing cross ventilation so necessary

Above: ***We found historical photos of the home from the rear yard.*** Right: ***The Pitzer family photo with Lee on the right, Blanche his wife to his left along with Clifford, and Robert on the pony. The girl holding the reins was Jean McBurney. The family dog, ignoring the whole affair, enjoyed the coolness of the stones.***

in the years before air-conditioning isolated us from this natural cooling, yet nature did intrude on this new rock home. Rattlesnakes found this courtyard a pleasant home, too, and Blanche, quickly ordered it to be enclosed.

Later, around 1920, Lee Pitzer took ownership of the ranch and the citrus groves. He changed acres of orange trees to lemon groves and soon after learned an expensive lesson about the extremes of Southern California's weather. Within a couple months during winter, the daily temperature could plunge from one-hundred degree heat to freezing.

Life during these early decades was inconvenienced and rudely challenged by the traumatic interruptions of

Front porch to Pitzer residence, notice the artful placement of the river rocks.

two world wars, one great recession, a few earthquakes, freezing temperatures, torrential rains, and floods. Yet, the settlers stayed.

Pitzer built his home with the river rocks found on and in his land, an alluvial fan at the base of San Antonio canyon. He also had the barn, water pump building, generator building, and garage needed to operate his acres of lemon groves built of smoothed granite river rocks. In her book on Claremont City, Judy Wright mentions that "Mr. Pitzer bought his property between $25 and $50 per acre, but he said it cost from $100 to $300 an acre to clear the stones."[2] Yet Pitzer persevered in the role "suitable to refined middle-class growers."[3] Promotional materials phrased this agrarian life of income and recuperation from the stresses of city life as 'Citrus Culture.'

Though local ranchers warned Pitzer that the ranch's land was too high an altitude for citrus to grow properly, within a few years his mature lemon groves contributed to the stratospheric growth of citrus production in the Pomona Valley—from 6,300 crates in 1892-1893 to 2,000,000 crates in 1916-1917. The invention of ice-cooled boxcars allowed citrus fruit to be shipped from the San Antonio Fruit Exchange in Claremont to eastern states with some shipped on to Europe. Citrus was a delicacy to those eastern snow-bound people, while in Southern California the sun and abundant water produced two to three crops a year.

Judy Wright gives us a rare reference to the men

A former owner of the Pitzer home in the 1980s, psychologist Patricia Hedges, describes her feelings for the home, "There's something solid and enduring about this place that makes us feel better." In an LA Times article by Mary Barber, on February 12, 1987, Hedges expressed an essential understanding that is shared by all the river rock homes owners we interviewed.

who built Lee Pitzer's stone buildings. "Victor Zuniga, Pedro Sevilla, and Juan Lopez as well as other men from the east barrio (in Claremont this was the neighborhood of the citrus workers) were skilled stone craftsmen."[4] History lost these stone masons, but it seems reasonable that many were of Mexican or Indian heritage, people who lived below the radar of history.

"A neighbor across the street said that we should have never moved in on Friday the 13," Lee said, "and that we would have bad luck...We sure did. In January 1923, the weather man dropped the temperature to 18 degrees...I ran water in the grove hoping it would raise the temperature some, but the water froze and I had a lake for the children to skate upon for several days."[5] Over five years were needed to repair the frost damage to the lemon groves, replanting and salvaging what they could to return the trees to production. Pitzer talked about water and Judy Wright shares his story on water technology then, "Water cost about $100 or more per acre, if you could find it...but I finally got a (water) witch to go over a spot where I wanted to put a well and sure enough we found some water."

The home's wood interiors of oak, walnut, and mahogany with the vaulted ceilings sheltered and protected the Pitzer family for fifty years. But its protective presence did not last for the Pitzer family. Blanche became ill, and she and Lee moved from their stone home to a retirement home recently built on lemon grove land that Lee had donated. He dedicated the sunset of their life to caring for his wife. Lee died in 1969, at the age of ninety-six.

The Pitzer home was placed on the National Registry of Historic Places on September 4, 1986 recognizing both Lee's contributions in both building the town and building this beautiful home. The Pitzer property now provides a quiet, calm environment for its current owner, Sunrise Assisted Living Community, who uses the stone home for small conferences and meetings.

Pitzer's story may not have been unique. These foothill settlers' lives had one element in common: barren, rocky land. Roads and utilities did not exist; their homes did not exist; and dry cleaners, gas stations, or grocery stores did not exist. With his brother, Russell, Lee kept alive the forward-thinking of those community builders, and among other civic projects, they founded the Claremont McKenna Colleges.

Though founding private colleges appears philanthropic, "colleges were 'great immigration bureaus'," Frank Walker mentions in his 1950 study of Southern

Before repurposing, this stone outbuilding was the home's garage. The photo may have been taken in the late 1980s.

After reconditioning to fit the needs of the new retirement community home. Photo taken in the 2010s.

Photos of the restored Pitzer Ranch pumphouse.

California, "and the newly platted community which could boast of a projected college had a distinct advantage over its neighbor."[6] Throughout the twentieth century these colleges have maintained an excellent standard of education for their intelligent and talented students, and today the Claremont colleges remain in high demand with stiff entrance standards, experienced professorial staff, and high tuition.

But these are facts culled from recorded history, not the 'why of what occurred'. History remains silent on why they built with river rock. But, we can surmise some possible 'whys'. Lee Pitzer was a wealthy businessman, not a farmer. His groves were a business. His neighbor and fellow settler, Dr. Johnson, had accumulated a good amount of money also. Both hired the same architect, Robert Orr, and this may not have been coincidental. In 1910, hiring an architect to design and build your home and ranch buildings was a luxury afforded only by the wealthy. Both Pitzer and Dr. Johnson were smart, educated, and visionary. They chose to build in stone, though wood was readily available and preferred by most. Stone's character told all who would see your buildings these qualities: lasting, memorable, immoveable, fire and weather

Close-up of the stunning and artistic placement of the colorful stones using tight mortar joints. This was the work of a master mason.

resistant, strong character, and affluent. Characteristics that could be said of their owners.

These homes were elegant in their use of a common rustic material. I believe these men planned on staying and Pitzer stayed for fifty years. Their foresight and acumen endures, exemplified by their stone buildings and the community they built, both thriving 110 years later. Possibly these men thought about their heritage as

This historic photo gives us an understanding of the abandoned state of another ranch outbuilding, possibly housing a well pump.

Reconditioned and repurposed, the old pump house has a new life for the community as an exercise facility. Both photos show the identical location of the doors, so we are looking at the same wall of the same building. A great coincidence to see these before and after photos.

An archival photo of a Pitzer farm barn at 926 Baseline Road, shown on the bottom and the ranch manager's home peeking from behind on the top right. Photo from Library of Congress. Prints and photographs online catalog, Reproduction number: HABS Cal, 19-CLAMO, and HABS 2267-D (Historic American Building Survey)1 D-4; Historic American Buildings Survey. Of the National Park Service, Department of the Interior.

Photo from 1987, LandmarkHunter.com. original photo from Library of Congress (LOC). The pump house before street widening

Pitzer Ranch Pump House, 240 feet west of Padua at Baseline Road. This uncut fieldstone pump house was constructed to irrigate an 80 acre lemon grove. "Fieldstone was used as a building material due to its availability and a need to clear the fields of fieldstones for agricultural use." National Records of Historical Places NRIS Number 86002192, Library of Congress. Photo from 1987

well as practicalities, responsible for aging men to want monuments to their extraordinary lives and character.

For those interested in preservation, the old Pitzer pump house received new life illustrating how the seemingly conflicting priorities of profit and preservation can coexist. The City of Claremont brings sound direction and language to the restoration of significant buildings, not just by telling us how it can be accomplished but, more critically, showing the interested and committed what shape and form preservation can be in the twenty-first century.

In the midst of the Citrus Glen residential development on the south side of Baseline at Padua Road, (Monte Vista Street) this stone building, built for the Pitzer ranch, breathes a renewed life. Driving down the development's private street to better view its neighbor, we were amazed to find the refurbished cobblestone barn located inches from Baseline Street in perfect condition. It felt like opening your front door and finding a new car in your driveway and your old clunker gone.

Repurposed for the residential community, this stone farm building now serves its residents as an exercising and meeting room, adding pride and history to their neighborhood. Can your stare leave that stone wall with its rich colors of orange, white, and grey of the thoughtfully placed river rocks? A visual delight lacking in the materials of current buildings. Note too the graded sizes of the stones with the larger ones being placed at

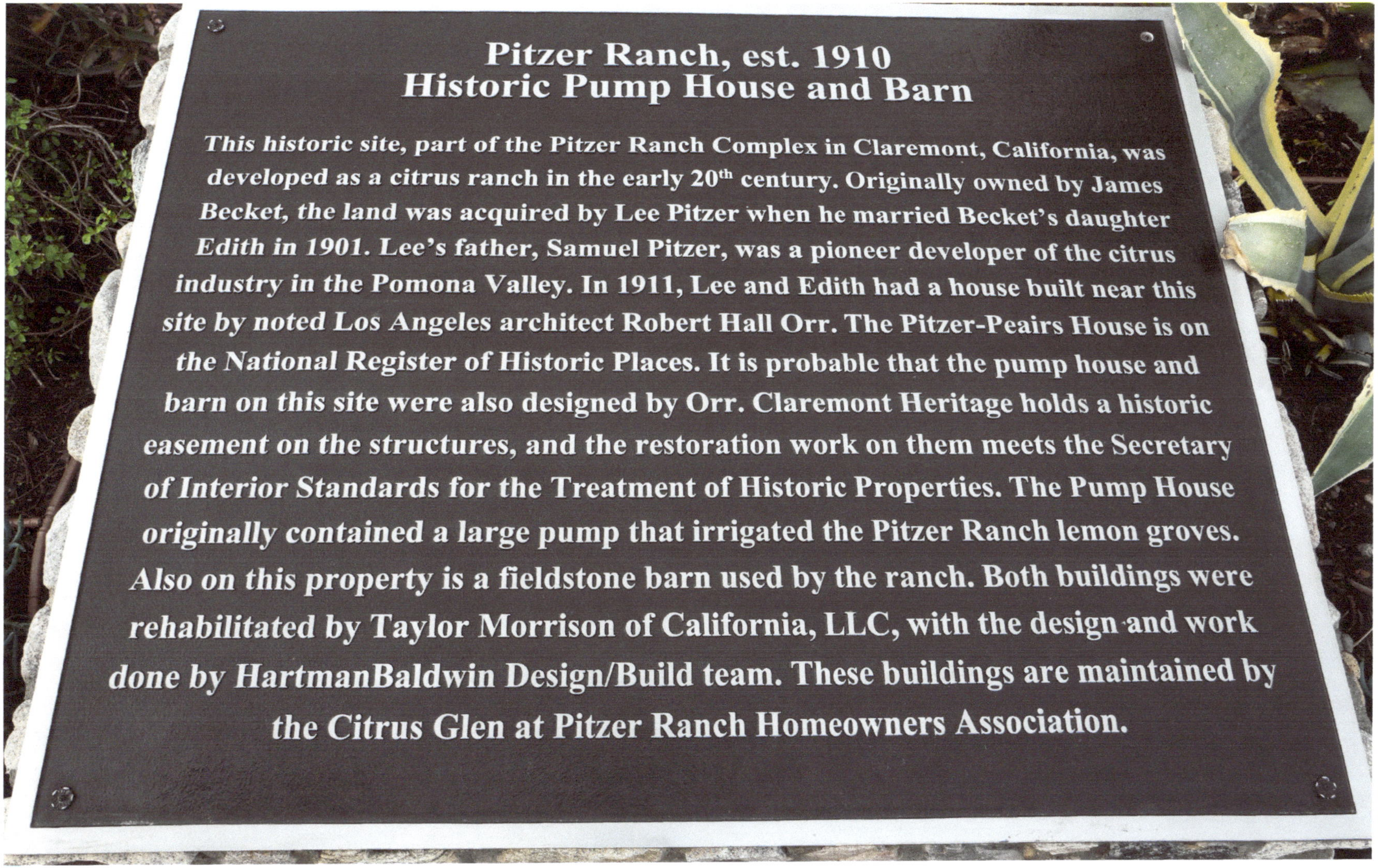

An aluminum plaque placed next to the base of the stone barn.

the base of the walls, the long thin stones used above the door and window openings, and the smallest stones marching along the top, the cornice, like a row of ducks. Delightful and entrancing. An outstanding example of Rockitecture.

Though these structures are utilitarian ranch outbuildings; the mason's natural artistic talent came alive when he placed the large stones at the top of the curved Mission Revival styled parapet. Today its refurbished condition is part of a condominium subdivision and an excellent example of restoring and reusing these historically significant buildings.

618 BASELINE

This building, constructed before 1915, served the Johnson Ranch complex. A century ago Baseline Street was a narrow dirt road and now has been widened to a

The ranch workers shared this stout residence, and the widening of Baseline Street removed a wagon/carport on this the north side of the building.

major four lane highway encroaching and threatening these precious buildings. Such insensitivity by institutions may ultimately doom these unique stone relics.

Constructed by Dr. Johnson's workers, the buildings show an experienced mason's talent. These fine examples of Rockitecture may also indicate Dr. Johnson's attention to detail and perfection that he learned as a surgeon.

1333 NORTH MOUNTAIN, DR. W. JOHNSON HOME.

In 1907, Dr. W. Johnson chose to use the river rocks, so prevalent in the area, to build his retirement home. Lee Pitzer and Dr. Johnson probably had become close friends for Pitzer built his rock home a few years later in 1913, using the same architect. The retired surgeon and doctor was obviously financially successful, for this home would be considered a mansion a century ago. The granite stones, collected from the adjacent washes and on the ground of his acreage, create a rustic contrast to the Georgian style white trim. The surrounding 168 acres of this rocky alluvial fan became orange groves creating an orchard that extended from Foothill Boulevard. to

"One of Claremont's fine stone homes," according to the Claremont Historical Society.

404 WEST FOOTHILL BOULEVARD (OLD HIGHWAY 66).

Sherwood Florist now occupies this restored landmark.

Based on the cars, this photo may have been taken in the early 1940s, and the occupant revised the company name to Sterling Supply Company.

The stone's swirls, white streaks, and black flecks occurred millennia ago when formed by underground volcanoes, liquified and blended together into granite blocks creating fascinating patterns inspirational to artists and designers.

Baseline Street and on both sides of Mountain Avenue. In the spring time, living among the sweet fragrance of the citrus flowers and the hum of the bees must have been a refreshing experience.

3232 Padua Avenue. Large two-story home with moderate pitch, cross gable tile roof. Bottom story is of uncoursed fieldstone and the top story is of stucco and stained, half-timber trim. The copper-hooded windows are interesting features of this house built in 1929. The house was built by Mr. and Mrs. Charles Schoolcraft, who picked many of the stones themselves from their lot and the surrounding area. Mr. Kidd and Ben de la Rosa were the stone masons. The green roof tiles are from Gladding-McBean.

Both its two stories of river stones and the steeply sloping roof are unique for river rock homes

3232 PADUA AVENUE

On a cloudy and drizzly day we stumbled across this magnificent home, set back from Padua Avenue in a front yard of old pines. A real estate listing (Redfin), described this home as having 3,574 square feet, four bedrooms, and four bathrooms. Built in 1929; bad timing for the Depression began later that year, Mr. Schoolcraft, the original owner, was quite a wealthy man for its two stories were unusual and this land was not expensive in the 1920s. It's possible that Mr. Schoolcraft preferred this design which may have been similar to his former home back east. Judy Wright wrote, "Mr. and Mrs. Schoolcraft picked many of the stones themselves from their lot,"[7] and the nearby wash to the east.

For this home there can never be too many stones. The site was bordered by these solid rock walls giving to all passersby a visually delightful work of masonry art.

Our recent photo, on the right, shows the building from the rear, the east side, while the historic photo, on the left, was taken from the from the front, the west side. This stone mansion remains in excellent condition, for which the owners can be rightfully proud. The roof retains the original ceramic green tiles, which are ninety years old, while the river rock walls remain sound without cracks and separations.

Left: ***Ninety years ago the Schoolcraft family planted their Christmas trees. For many years now these mature trees of joy and memories have provided cool shade over the expansive yards and spacious mansion.***

RUSSIAN VILLAGE DISTRICT

Cars speed by; the drivers focused on their destinations oblivious to these fifteen unique eighty-year-old masonry homes. A block-long amalgam of handmade stone and broken concrete homes. Konstanty Stys and friends built Russian Village during the Depression years of the 1930s. Immigrated from Poland, not Russia as some neighbors thought. Stys started life in the U.S. as a steel factory worker in Ohio during the early years of American industrialization. In 1921 his wife and family settled in Claremont where he worked on the construction of the Claremont Colleges, which the Pitzer brothers helped establish. Later he bought land on Mills Avenue, and in 1922 built his home of scrap wood from the campus construction site and river rocks from the land. He must have also bought or scrounged left over Portland cement necessary for the mortar.

333 South Mills Avenue

306 South Mills Avenue

MILLS AVENUE, RUSSIAN VILLAGE

Unlike the Stonehurst village in San Fernando Valley, which we'll visit in chapter 14, where all the rock homes have the same plan and exterior design, Stys built the Russian Village homes of material salvaged from building demolition sites, construction sites, wrecking yards, and by making friends in the City's Public Works Department to obtain excess and discarded materials resulting in each home being unique. Stys encouraged relatives to join him and relocate from Eastern Europe to Claremont. He sold them lots cheaply, around $300, and helped build their new homes.

Broken sidewalk concrete formed the exterior of a couple homes, while mortared river rock formed the walls of others. All the homes used clay mission tile roofs, again salvaged from demolished schools or public buildings.

336 South Mills, Konstanty Stys and friends completed the laborious days-long effort of gathering the cobblestones from nearby citrus groves and the adjacent wash below San Antonio canyon.

316 South Mills Avenue

360 South Mills, built in 1934 during the decade of Depression, an economic hangover from the party decade of the 1920s. As with all these rock homes, the mason chose his fieldstones with care for size and color "with the largest stones placed at the base."[8]

The 1933 Long Beach earthquake proved that one man's loss could become another man's gain, for Stys that disaster provided an abundance of building material.

An early settler on the edge of Claremont amidst the lemon groves, he created a small community of like-minded people wanting to live their dreams. Responding to tough economic conditions he showed a rare gesture of social cohesion, generosity, and ingenious artistic talent.

During the 1930s, these fifteen homes cost a modest few hundred dollars in materials. Now they sell for hundreds of thousands to owners who cherish their special homes and have successfully protected the neighborhood from destruction, from street widening, or rezoning while getting them listed in the National Register of Historic Places.

EAST PALMELLO STREET

This perfect example of craftsman style stone home seeks a compassionate owner, with the ability to restore its charm. Sitting immediately adjacent to both the 210 freeway and Indian Hill Avenue, it's ideally situated for a professional office.

On the north side of Palmello Street and located among expensive modern residences, is this delicate stone building constructed by the local telephone company recently to house its telephone switching equipment. Using river rock to resemble Dr. Johnson's and

932 East Palmello

900 block of East Palmello Street

A historic jewel in a town of jewels. The two circular penetrations above the door and window may have been for the exhaust pipes of the diesel generators that supplied the electricity for the water pump.

West side of the Mira Mar home with the rich rust toned river rocks of Claremont.

Lee Pitzer's original river rock buildings, it was built in the 1980s and shows the importance the City of Claremont gives to retaining its heritage.

Built in 1913, this river rock building has remained an active ground water pump station. The curved parapets in the Mission Revival style popular then illustrates that these Claremont settlers gave attention to appearances and details. With careful effort to maintain its original design, the city used the original blue paint and maintained the rusty tin roof.

1108 E. MIRA MAR

For decades Nelson and his brother David have proudly maintained their family home in its original condition. The lot is large, over an acre and located on the alluvial fan close to the mountains. Nelson's family home comes with an unusually large three-car stone garage. Typically in the 1930s, garages were built for one car, if built at all. The home is also larger than average with its 1,900 square feet providing spaciousness to its two bedrooms and living areas.

Three-car river rock garage, a rare sight in the 1930s

This stone home rises naturally out of the stone patio and stairways. One with the land, it transitions seamlessly from the stone patio.

808 Baseline. Built in 1922 probably for a family, when Baseline was a dirt road, with its size of 2,300 sq. ft and four bedrooms and two bathrooms. Some of this may be additions but we could not tell from foliage blocking the home at the rear.

While we toured his extensive gardens, Nelson spoke lovingly with deep pride of living in this family home. The street's name—Miramar—is Spanish for 'look at the sea'. Though forty-six miles distant, the Pacific Ocean may have been visible on clear days, but now that view has been obscured over the years by trees and buildings and city haze. In a large wash one lot to the east, we find the source of these colorful river rocks.

Aside from the California Missions, these Claremont homes are among the oldest continuously-occupied California structures, and they are built with river rock. Many earthquakes have shaken this area for the past century. Many are mild, but the repetitive shaking can add up. Yet none of these buildings were cracked or

4404 North Towne. The recent carport addition blocks the view of the stone home behind it.

damaged, their owners not noticing any damage nor were they concerned about their safety in their homes.

That afternoon we traveled eastward into the towns of Upland and Rancho Cucamonga, where the Rockitecture homes in these towns were built from quite a different lineage.

END NOTES

1 Pomeroy, E., *Lost and Found II*, Many Moons Press, 2002, p. 122.

2 Wright, J., *Claremont: A Pictorial History*, Claremont Historical Resources, 1999, p. 140.

3 Starr, K., *Inventing the Dream*, Oxford University Press,

1985, p. 143.

4 Wright, J., *Claremont: A Pictorial History*, Claremont Historical Resources, 1999, p. 154.

5 Ibid., p. 140.

6 Walker, F., *A Literary History of Southern California*, University of California Press, 1950, p. 160.

7 Wright, J., *Claremont: A Pictorial History*, Claremont Historical Resources, 1999, p. 309.

8 "Walking Tour of Russian Village," pamphlet, Claremont Historical Resources, 1981.

CHAPTER 13

Upland

"Gentle Grade, it is, literally, the Upland of the gentle grade which begins west of Pomona and rises through that city and Claremont, growing gradually higher until at Upland, it approaches the foothills."[1]

The Bethany Union Church around 1920 with a line of power poles providing locally generated electricity. Upland had lots of barren land then.

The settlers named the sloping lands of the foothills, not ingeniously, "North Upland." Later, in 1891 this land on a large alluvial fan at the base of Mt. San Antonio, the tallest peak in the San Gabriel Mountains, became the City of Upland. Here in the eastern part of the San Gabriel Mountains, from canyons with names like Bear Canyon, San Antonio Canyon, Ice House Canyon, the granite cobblestones were born and found.

In Upland these stone homes began appearing in the early 1900s, when people of wealth either retired here or prospered by planting their ranches with citrus groves.

A proud owner of his new 1921 Model T Ford

1060 W. Foothill Boulevard, Upland.

1060 WEST FOOTHILL BOULEVARD

In the late 1800s, George Augustus Hanson (1867–1941) immigrated to Upland from Canada and developed his prosperous G. A. Hanson Fruit Company. The fruit whe grew and sold was either lemons or oranges. In 1896, Hanson, 30 years old, hired masons to build his family home of stone and this house still exists today, though it has been repurposed as an office while being restored as a historical monument.

Hanson planted bare land to raise and sell citrus and hired masons to build a river rock home and ranch buildings. Later with his brother, George, they applied their practical minds to inventing and patenting. During a hot day in July of 1918, they perfected a process of sterilizing and coating oranges to preserve freshness

One of the first stone homes in Upland, no longer existing.

while being shipped to the east. Much later another set of Upland brothers, named the McDonald brothers, would invent the industrialization of fast food and change the nation's eating habits like the availability of fresh oranges changed the consumption of fruit in the U. S.

In the early 1900s many cobblestone buildings were built to support the citrus industry, the second largest industry in California. "Between 1880 and 1920 California recreated itself into an agricultural empire," the historian Kevin Starr writes and "citrus and viticulture led the way."[2] California was no longer the wild west of the gold rush years.

Though Claremont and Upland were the center of citrus groves, Ontario, just south of Upland, became the land of vast vineyards, decades before Napa and

George Hanson's family and their stone home which remains on Foothill Boulevard.

Sonoma became famous for their wines.

These enterprising ranchers and businessmen not only used the rocks of the land to provide enduring buildings, they unknowingly provided a legacy of their

The restoration of this generator building began in 2020 refurbishing it for commercial uses.

work by creating new towns. Industrious and wealthy people like Pitzer in Claremont or Dr. Johnson hired stone masons and within five years constructed more than ten river rock structures for operating their vast acres of citrus groves along with their private residences. Their masons' labor involved finding, hauling, sorting, lifting, and placing about five stones per square foot of wall weighing on average 150 pounds per cubic foot.

These stone walls vary in thickness and slope outward from a foot at the top to two and a half feet thick at the base, with a section of wall one foot high by one foot wide weighing around one hundred and sixty-six pounds, more than the weight of the old upright pianos popular in those days. Lifting and carrying each cobblestone weighing thirty to fifty pounds up to seven times to place it in the wall required strong resilient men. The stone walls

Top: ***This scene shows that both the men and women harvested the vineyards during the late 1800s. Taken in the Ontario area the San Gabriel Mountains provide a monumental and scenic background.*** Top Middle: ***A field of oranges ready for packing to ship by rail to the eastern states.*** Above Right: ***Upland 1936 Citrus groves dominated the land use then.*** Above: ***Bob Black, a teen in this photo, guided the horses to weed a row between the orange trees.*** Right: ***Citrus packing house workers in early 1900s.***

Photos on this page from online archives of California, contributed by Upland Public Library.

25 West 17th Street, Upland

The stone garage built in the rear of the two-story mansion. The apartment above the garage housed a servant and family.

1790 North Euclid

Above: *This photo exemplifies the high level of wealth that settled in early Upland as an architect designed this stone-faced home in 1920 for the Howe family. This expansive 3,580 square foot California stone home included four bedrooms and two bathrooms, when the typical modest stone home was 700 to 900 square feet.*

Left: *Considered a mansion when built in 1905 and renovated in 1978, this home includes 4,200 square feet with three bedrooms and three bathrooms.*

4436 North Towne
A cobblestone home for the citrus ranch supervisor and family. It sits next to an abandoned cistern that once stored water to irrigate the citrus trees.

2393 North Mountain
These stone buildings were the carriage house, then garage and residence for servants at a large wood-framed home not shown, We visited with Robert (Bob) the current owner who with generosity and politeness shared with us a little of the home's history, before it burned. The original home owners built in upper Upland in 1906 amongst their large citrus ranch. This estate later burned in 1988 and Bob rebuilt the manse to a detailed duplication of the original, such is his devotion to the home's heritage.

HISTORICAL CONTEXT

These two historical photos provide context to the culture and living conditions in those early years of settlement.

Framed with vertical wood studs that supported the horizontal shiplap siding on the exterior, Mr. George Barlow's modest home shows the preferred house of the settlers in the 1890s that featured a nod to Victorian style with detailed woodwork in the lattice and the curved arch above the steps. In the front yard appears to be an electrical pole. The women of the family probably dressed in their finest full skirts and white blouses for this rare photograph.

The interior of a modest home where the resident sat in a room which performed many functions, such as kitchen, dining, library. A close look shows utensils, books, knickknacks and the sheet metal chimney for the cast-iron stove that provided the heating and cooking.

On the north side of the granite stone bungalows is the narrow driveway leading to the rear parking area. When this was built automobiles were not owned by everyone and the center strip of grass absorbed the oil which dripped from the early cars.

581 North Euclid
Three stone buildings are arranged around the central walk court. Two are one story with the walkway ending at a two-story building. Per one of the tenants, the building housed grove workers while the two-story section housed the supervisor and his family.

The citrus orchard workers of Upland found the simple housing they needed in these court-side stone bungalows. Today its residents find it magical to live in one of these seven homes—simple, quiet, rugged, and one of a kind. In the 1920s, when they were built, the grove foreman lived in the two-story unit at the rear of the lot while the workers moved into the 600 square foot single story bungalows.

Ontario and San Antonio Heights Railroad Waiting Station in early 1900s. Topping this modest structure the crenelation appears to have been added for the stone joint about two feet below the top of the parapet is in a straight line, which never occurs in the wall area, but only at the top of the parapet.

in the modest home of six hundred square feet could weigh up to one hundred and eighty thousands pounds or ninety tons of stones, the same weight as an adult Orca whale. The weight of the same sized wood house would be less than half that of a stone home. Building with stones and mortar required extensive labor; building with lighter weight wood, not so much.

SAN ANTONIO HEIGHTS WAITING STATION MOUNTAIN AVENUE AND 24TH STREET

In the 1880s you'd wait for the donkey-powered trolley in this stone waiting station at the top of Mountain Ave. It took passengers downhill to Ontario's commercial district, about a six-and-a-half mile ride from this wealthy

San Antonio Heights neighborhood. A short distance further up Mountain Avenue lies the ruins of the power station that provided electricity for the trolley and the local residences. But before being electrified, two mules pulled the trolley uphill and rode it downhill in an attached trailer as gravity rolled the trolley and donkey trailer down to the city.

The two donkeys where no longer needed once electrical motors operated the trolley and were sold to a local farmer who had thought he got a good deal. As the story is told, the farmer wanted to return the donkeys to get his money back, the donkeys refused to move. Then someone mentioned, probably a conductor, that he might also want the bell that signaled for them to pull. The farmer attached it to his plow, rang it, and the donkeys moved–until the end of the row being plowed. There he tried to get the donkeys to turn around and go down the next row to be plowed. They stopped. He tried and tried and then remembered how they responded to the bell; he realized the donkeys were waiting to climb onto a trailer to ride back to the beginning. It took some patience and time for the donkeys to walk down the row of crops without climbing on a trailer for a free ride.

Trolley tracks in the middle of Euclid Avenue which ended in the foothills of Upland.

Today, the donkeys are long gone and the waiting station rests in the San Antonio Park. A couple blocks from this delicate passenger waiting station remains the stone Ontario Electrical Power House. The growing popularity of automobiles retired this trolley system around 1915, and in 1925 the station and the adjacent fifteen acres became

Stone Castle powerhouse looms through decades of foliage.

a public park. In September of 2012 the National Register of Historic Places included this waiting station as having special significance.

Researching and interviewing came to a halt, as mentioned, due to COVID restrictions. and one puzzle remains: Why build the waiting and electrical stations on Mountain Avenue when the trolley was located on Euclid Avenue?

STONE CASTLE POWERHOUSE
2453 NORTH MOUNTAIN AVENUE

Decades of growth cover this roofless fortress. It appears from under entangled vines, like a kelp-covered monster emerging from the sea. Once this stone castle housed generators to supply electrical power to the steep Euclid Avenue passenger railway and the San Antonio Heights neighborhood. Blanketed by a century of nature's growth, this fortress masterpiece may strike fear and wonder in all but the intrepid. Today nature transforms the stone castle into spooky ruins possessing the haunting elements of a Halloween set that E. A. Poe would admire. Mysterious, it was once a vital source of electricity but is now disused and abandoned letting nature erode this stone wonder into a fascinating architectural ruin.

Atlas Obscura, a website featuring unusual locations throughout the world (www.atlasobscura.com) describes the Stone Castle Powerhouse "as a testament to time when architecture was beautiful and not just utilitarian."[3]

Built by Ontario Electric Company, it operated a short eight years from 1887-1895 then a fire in 1915 resulted in its abandonment. "Thankfully," *Atlas Obscura* said, "because of its stone construction and its protection (in 1975) by the National Historic Register and a healthy ring of poison oak, these ruins may be around, not covered by graffiti for generations to come."

END NOTES

1 Upland Heritage, www.uplandheritage,org, viewed January 2, 2020.

2 Starr, K., *Inventing the Dream*, Oxford University Press, 1985, p. 134.

3 Gage, J, "Stone Castle Powerhouse," *Atlas Obscura*.

CHAPTER 14

Rancho Cucamonga, Riverside, Irwindale, and Ontario Cities

Rancho Cucamonga

"It was a wonderful life. I have tears in my eyes thinking about it, it was really great."

Rancho Cucamonga Oral History in 2001 of Peter Tolstoy on growing up at their citrus ranch in the 1920s and 1930s.

Many foothill cities prospered with the financial successes of the citrus groves. Located primarily in the San Bernardino and Riverside Counties, which are about thirty miles east of downtown Los Angeles, wealthy businessmen from the Eastern U. S. planted and irrigated tens of thousands of acres of citrus groves, developing this area into a major economic powerhouse. And many of those businessmen built their homes and ranch buildings with the river rocks of the land.

Seven centuries before them the Kukamongan Indians thrived on this alluvial fan below the 'Cucamonga Canyon'. Kukamonga, the 'sandy place', the Indians named this land where later white settlers would transform it into untold acres of irrigated vineyards and citrus groves.

Located below Cajon Pass, the only break in the sixty mile long San Gabriel Mountain range that is caused by the San Andreas fault crossing through it. Rancho Cucamonga became a natural center where transportation routes met and crossed connecting many parts of California. Here was the old Indian trading route, the famous Mojave Trail, which later became the Santa Fe Trail and Butterfield Stagecoach Route; with the arrival of cars it was renamed again as Route 66, the Mother Road,

The Kukamonga Hunting Grounds. The foothills provided deer, atelope, rabbit, quail, and coyote for food and clothing. The chaparral plants provided herbs, food, medicines, and basket-making materials.

and now Highway 15. These old roads continued along the foothills connecting towns west such as Los Angeles, Long Beach, and the ocean ports where Mojave Indians traded with the Chumash and now the U. S. trades with nations throughout the world.

DEMENS-TOLSTOY HOUSE, 1890
9686 HILLSIDE AVENUE

The plaque says, "Once part of a large citrus ranch, this house was the home of Captain Peter A.

Demens. A Russian nobleman instrumental in bringing the railroad to Alta Loma...Demens was also an accomplished journalist, railroad and lumber magnate, and co-founder of St. Peters-berg, Florida." The plaque was placed during 1994 in memory of Captain Peter and Raissa Demens and Andrey and Ann Demens Tolstoy. A close look at the large home reveals that the first floor is river rock while the second floor is wood framed. All the small stones indicate the construction may be veneered stone on wood frame support. Though gated, this beautiful mansion can be admired from the street.

Pyotr Dementyev (1850-1919) emigrated from a wealthy family in Russia. An educated aristocrat, he reportedly grew up in a stone mansion maintained by servants. An uncle of Christian nobility raised Pyotr tutoring him in the management of landed Russian estates. This provided the skills for his eventual purchasing a large citrus grove. Emigrating to the United States in 1885—or maybe Pyotr had been exiled from Russia for his outspoken criticism of the dictatorial Czarist regime—his Russian name became anglicized to Peter Demens. An energetic young man in his thirties, he first stayed with his uncle at his uncle's orange groves in Jacksonville, Florida where became more familiar with grove management.

From there he entered the forestry business, which he learned from his Russian uncle, and sold railroad ties which he parlayed into owning a local railroad company when that company couldn't pay.

After expanding the railroad company, he sold it in 1889. But a year before then, Demens, ever the restless entrepreneur, founded, along with a partner, a new town in Florida. Of course he named it after his hometown in Russia, St. Petersburg.

Moving to California in 1895, Demens bought a home in Alta Loma from the Chaffey family. The Demens family was large so filled the twenty-seven room estate called the 'Stone Castle.' (Apparently the Chaffey family member's new wife refused to live in Alta Loma, so the estate and groves went up for sale.) While maintaining his groves Demens wrote to friends in Russia to encourage and attract Christian Russian groups to join him. Many did and many became employees. Demens retired here and during that time his youngest of seven children, Vera Demens married into the famous Tolstoy family.[1]

THE ALDEFER HOME
8308 BAKER AVE.

Though built in 1927 this home looks recent, certainly not about to celebrate its one hundredth birthday. Subsequent owners of this 2,700-square-foot family home of four bedrooms and two bathrooms have maintained it in like-new condition. When built, orange groves must have surrounded the home, along with a dusty unpaved road. In those early years city noise did not disturb, blue skies were enjoyed daily, and rains were infrequent, but the home's steep sloping roof indicates that the owner or builder, perhaps both, immigrated from wet and snowy climes. The short overhangs lack the necessary Southern California need to shade the interiors from the intense sun. Promotional literature bragged to the cloudy and grey skied citizens of the eastern states that the sun shone 287 days a year.

A view of the back entrance of the United Methodist Church.

UNITED METHODIST CHURCH 7900 ARCHIBALD AVE., RANCHO CUCAMONGA

'The Old Stone Church of Cucamonga' demanded to be constructed. When neighbors of Cucamonga (renamed Rancho Cucamonga) thought a church was needed for their religious services and social events, they built a wooden church. It may have been similar heavy winds which blew down the wood church in Sierra Madre which also destroyed the Methodist neighbors first attempt at church-building. Committed, they rebuilt. Also of wood. It too blew down. Undaunted, these determined parishioners built a third church in 1907, and as with our Sierra Madre stone church, this third and last church switched to stone construction reinforced

The United Methodist Church, Rancho Cucamonga

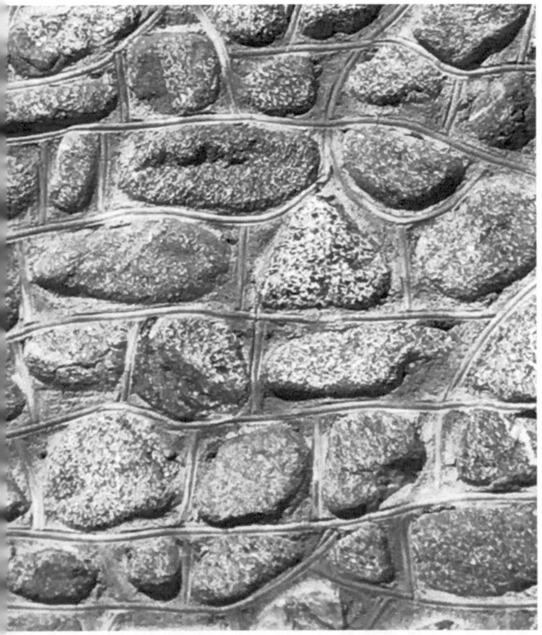

with wire in the backing mortar. Should telephones have been installed then, these Cucamonga folks could have learned from their Sierra Madre neighbors twenty miles to the west of the winds we now know as 'The Santa Anas' or 'Santanas' or 'devil winds'. Forty-mile-per-hour hurricane force winds roar down the mountain canyons each autumn, not only damaging and demolishing buildings but whose force and dry air scour the foothills starting fires in the oily and dry mountain chaparral. Then the winter rains erode the soil causing large and fast moving mudflows, and yet more destruction.

These determined community builders would have their church in spite of those ferocious desert winds. Three hundred and thirty-seven horse drawn wagon loads of stone and $7,000 later, in 1907 they celebrated their new stone church. The river rocks though were not lying around like in other areas. The stones were found up the steep foothills, carried, loaded, and wagoned down from the Cucamonga Canyon five miles north. Looking now at this stone church we need to understand that for each of these three hundred and thirty-seven wagon loads of stones, resilient men lifted each stone at least five times. A strong testament that many of the stone homes and churches were constructed not because the 'stones

were there' but because of the practical need to resist strong natural forces.

Five generations later, after many wind storms and earthquakes, the Methodist community still enjoys its stone church.

9953 HILLSIDE, ALTA LOMA
AT EVENING CANYON AND HILLSIDE

Built on 'high land', the thoughtful placement of the cobblestones is seen at the top of the windows with the specially selected long, thin stones forming the arches. A uniquely rich texture of varying greys, sizes, and orientation, the stones excite our eyes dulled from constantly seeing stucco's blandness.

Again, the steep roof slope is so foreign to Southern California's climatic needs and historical styles; yet it remained familiar and preferred to the early settlers from the snow country in the eastern U.S.

Cherbak's family home of 1913 shows no aging with its excellent maintenance, a home well worth viewing.

THE A. P. CHERBAK FAMILY HOME 9983 HILLSIDE, ALTA LOMA

Walk west a couple of lots and you'll find this stone home built in 1913. You'll notice that these homes are of substantial size, which indicates the wealth of their first owners. This home has 2,880 square feet with six bedrooms and three bathrooms; its freestanding claw tubs and the dark stained, polished hardwood floors retain its original era. Details of the stone's careful placement indicates the same masons may have built both homes, especially since they are neighbors. One hundred yards to the west lies Crooked Creek Wash and north a few hundred yards are Deer and Day Creeks, the sources

of the river rocks. Cherbak responded to his neighbor's advice, Peter Demens, and Cherbak emigrated from Russia to this Alta Loma area. Here he built this sixteen room family home for his six sons. But Cherbak's love for his country resulted in him in returning to Russia.

Earlier we learned that Frenchman Le Mesanger longed for his native home, returned to France, and later died in his hometown; Cherbak also left his family to return to his native country, Russia, with patriotic zeal to help establish a revised government after the first Russian revolution in 1905. "The new regime," said the book *Rancho Cucamonga*, "did not allow Cherbak to return to America"[2] never to return to his family home in Alta Loma.

The Cherbaks built several stone homes and this one at 5510 Hermosa is known as the 'Cherbak House'. The six sons hauled the stones, which they called "Cucamonga potatoes," from the local canyons and washes. A modest home of 1,700 square feet built in 1930, it contains two

Above left: ***This restored Porsche speedster contrasts elegantly with the restored stone home, at 5510 Hermosa, the Cherbock House, eliciting the juxtaposition of speed next to solid stability. This model Porsche is the type that James Dean died in a crash on Highway 46, east of Paso Robles.***

bedrooms and one bath and is located on the same Hillside Road as the previous two homes. But this modest one, built fifteen years after the two larger homes, could have been for a retired couple from Russia to enjoy the dry warmth of Rancho Cucamonga in response to Peter Demens' encouragement.

RIVERSIDE

"There are few cities more beautiful than Riverside."[3]

The new town of Riverside, located next to the Santa Ana River, attracted wealthy settlers, their influence and capital bringing the two railways through this and the surrounding towns. Lemons and oranges were shipped to the east and tourists and immigrants from there were shipped west by both the Southern Pacific, which traveled the southern route through San Diego onto New Orleans, and the Santa Fe railroad company, which connected to Santa Fe, New Mexico, Kansas City, and then Illinois.

The Riverside city website mentions that its founding occurred in 1870 "by a group of Easterners who wished to establish a colony dedicated to furthering education and culture."[4] And the second California gold rush began, growing golden oranges on this alluvial soil under legendary sunshine.

Riverside prospered and by 1882 its settlers

Peace Tower and Bridge on Mt. Rubidoux, Riverside (Though these granite stones show their cutting and shaping, not our natural rounded river rocks, this memorial is included for its special significance.)

On top of Mt. Rubidoux shows the clear expansive view of the San Gabriel Mountains beyond downtown Riverside.

had established groves of more than 500,000 citrus trees. The fledging town became "the wealthiest city per capita by 1895."[5] That census probably did not include the many Mexican and Chinese laborers that harvested the fruit.

Orange grove country. A century ago you'd smell the spring orange blossoms, feel the intense sun in summer, and float in the clearest of blue skies, every day. Oranges were the capital of Southern California in 1900. "When orange crops fail," a note in the 2009 Journal of the Riverside Historical Society states, "as it frequently did, whole resources...are cut off and hard times are inevitable."[6] Such was the dependence on orange groves to Riverside and its adjacent towns.

World War I destroyed many European cities, empires collapsed, and millions died. Many of those millions came from the United States. Soldiers slaughtered in trench and gas warfare. WWI inaugurated the U.S.'s involvement in European wars. A second was to occur.

The citizens of Riverside built this Peace Tower and the adjacent stone bridge leading to the Tower on

Mt. Rubidoux Peace Tower, my mother is on the bridge, photo taken by my dad in the mid-1930s.

The cloudy sky on the day we visited the stone memorial unwittingly heightened its somber presence, brooding for decades in memory of the thousands of American soldiers who died in WWI.

the highest hill in the city, Mt. Rubidoux, to honor the American soldiers killed in that "War to end all Wars." On Sunday, December 13, 1925, thousands of Riverside citizens, enjoying a clear cool day of 50 degrees, hiked 30 minutes to the hilltop, a slight breeze refreshing them on their pilgrimage, to the dedication of this new stone memorial. When arriving at its peak they were rewarded with an expansive view of Riverside County covered with groves and vineyards that extended miles north stopped only by the high steep San Gabriel Mountains.

GUASTI VINEYARDS AND WINERY

Guasti vineyards and winery was a legendary operation and town named after the Italian who immigrated here then ultimately planted thousands of acres of vineyards in San Bernardino County. Secundo Guasti extended his special talent of encouraging Italians to also immigrate to the U.S. and join him at his ever-expanding vineyards.

In 1900, on cheap, dry, chaparral-covered land he purchased 2,000 acres from Mr. Signorio and ultimately enlarged the vineyards to 5,700 continuous acres using the labor of many Italians recruited from his home country, many arriving with families. During those decades a new 'company' town was established to include employee housing, a school, fire station, a modest chapel that continues to provide services to its congregation today,

Rock winery buildings at Guasti Vineyards, fenced and waiting for restoration.

and a post office that serves the community of Guasti. At its height around 1,200 Italians and Mexican workers lived and worked in Guasti.

Secundo Guasti immigrated to the U.S. in 1884. He had turned twenty-five then and settled forty miles east of downtown Los Angeles on the sandy soils near Cucamonga where he began his vineyards. Having grown up in the Italian Piedmont district vineyard, he knew the soils beneficial for growing grapes. And with sixteen other Italian immigrants, they pooled their savings and bought 2,000 acres for 75 cents an acre or for $1,500. In 2020 dollars that would be about $45,000. Not bad for immigrants beginning a new life in a new land. Their company, The Italian Vineyard Company, established their claim to Southern California Italian sweet wines.

Built around 1904, these large stone winery buildings housed the processing of the grapes, stored the grape juice during fermentation, to be bottled here. One of

Guasti Vineyards and Winery.

the first totally vertical operations in Southern California. Yet a centuries old practice in the 'old-countries'. During the Prohibition years his winery continued "operation as if Prohibition applied to everyone but them." As all of Congress' laws include them, Prohibition had exceptions to: okay to make wine for medicinal purposes and religious uses. As with many other vintners, these two exceptions allowed Guasti's winery to flourish.

Having established a vineyard, a winery, a community, and a large estate, the Italian immigrant died at sixty-eight and his family interred him in a stone crypt at Calvary Cemetery in east Los Angeles. From these large stone buildings in Ontario, used to produce and store wine, to a small stone crypt in east L.A., the Guasti heritage remains a vibrant story in the area's history. Currently the stone winery is protected by chain link fence showing that these buildings are not only special but sacred to the area's history and heritage.

Stone buildings were essential for keeping the wine barrels cool during the fermenting stage, crucial to successful wine. In other areas caves may have been dug; here in flat alluvial planes the abundance of cobblestones provided the solution to constant coolness. Notice the

2309 3rd Street, La Verne

different parapet designs; these were probably built by different masons years apart, one is simple gable roof, the next a unique style of simple peak sloping to match the roof, and the third a mission style with a flourish. All creative and worth preserving.

The site being fenced and protected from vandalism suggests the dedicated intent to refurbish these vacant buildings, shore up their sagging roofs, and reuse them. This is private property and a good viewing location is where I stood, west of Turner on the south side of Guasti town just north of the railroad tracks. Here stands the original Catholic Church still in use for the Italian immigrants and Mexican people who worked the vineyards and whose children remained in the area.

While visiting, a mile-long freight train with four engines rumbled by at a modest speed of 20 mph on century-old tracks, while a hundred yards north on the 10 freeway, motorcycles scream by at high speeds. These timeless stone winery barns remain unfazed by both the ground vibrations and high pitched whines, their covered windows deaf to the noise and blind to the setting sun.

We follow that train and now drive west to North Hollywood, and the San Fernando Valley, another site of California's first vineyards.

OUR LADY OF GUADALUPE MISSION CHURCH
16239 ARROW HIGHWAY, IRWINDALE

On March 4, 1990 the *Los Angeles Times* published an article by Irene Chang profiling this iconic historic church. A few years earlier the Catholic archdiocese wanted to build a new larger church to handle the congregation growth since 1917 when Our Lady was built. The city of Irwindale didn't allow the demolition of this historic beauty, eliciting much applause for the city

council people in office then. The city bought the stone church from the archdiocese. On our visit a couple years ago the church was closed with no one around. The city might be raising funds to refurbish the building for civic use, such as home to a historical society, or community center, or arts center.

On the side is the stained glass window of Our Lady of Guadalupe

The *LA Times* article described the church and its 1990 situation: "After lengthy negotiations with the Archdiocese of Los Angeles, the City Council unanimously voted February 22 to purchase Our Lady of Guadalupe Mission, and to make costly earthquake renovations to the structure. 'This is so the future generations will know their forefathers built it,' Councilman Patricio Miranda said...The 100-by-200-foot mission, dwarfed over the years by development and redevelopment along Arrow Highway, was closed by the archdiocese in 1988, after engineers determined that its unreinforced walls might not withstand a major earthquake. The archdiocese said it could not afford to spend $100,000 on steel reinforcements and other structural additions to the mission."

The home for the former pastors, located at the rear of the Mission Church

What is interesting here is that if the archdiocese could not afford that hundred grand, how did it afford to build the multi-million dollar church of the same name nearby on Cypress Street? And did the engineer who proclaimed a death knell for this beauty research how this cobblestone structure survive earthquakes for seven decades? But that harkens back to the earlier chapters on earthquakes.

"It sustained minimal damage in the 1987 Whittier Narrows quake, showing only a few faint cracks in the stone walls," said Father Patrick M. Hughes, pastor of Our Lady of Guadalupe parish. A common observation and sentiment of all cobblestone structures, illustrating once again that practical knowledge at the construction site of how to build frequently exceeds theoretical structural calculation on paper.

"Women made tamales and tostadas for the workers," according to a booklet detailing the mission's history, written in 1969 by Father Anthony Z. Marigo, the church's first full-time priest. "Children sold raffle tickets to buy cement for the foundation. And in 1917, when Our Lady of Guadalupe Mission was completed, the first Mass was celebrated using a rough, handmade table as an altar and linens and candles borrowed from the nearby Azusa parish."

"In 1965, the archdiocese built a 700-seat Our Lady of Guadalupe Church at Cypress Street and Irwindale Avenue to replace the 90-seat mission, which was then relegated to ceremonial use." Though the City now owns the building and the land the residents and parishioners own fond memories from decades of family use. "Even after it passes into city hands, some residents say they still want to have a say in the mission's upkeep.

"We maintained it," said Theresa Chico, 38, president of the Guadalupanas, a group of parishioners who have taken turns cleaning and fixing up the mission over the years.

"It's the people's church," she added.

END NOTES

1 Emick, P., *Rancho Cucamonga*, Arcadia Publications, 2011.

2 Ibid.

3 Clymer, D.H., personal correspondence, Riverside Historical Society, 1895.

4 City of Riverside, riverside.gov, accessed June 26, 2020.

5 Ibid.

6 *Journal of Riverside Historical Society*, February 2009.

CHAPTER 15

North Hollywood, the City of San Fernando, San Fernando Valley, and Stonehurst

"Waves curl over the beach at Nordoff Street • In the soon-to-be San Fernando Valley • Sand and river rounded stones."

"The San Gabriels are battered and crumbling from mountain building earthquakes, and mass erosion. The Santa Ana winds and chaparral fires on steep slopes followed by spring rains bring catastrophic mudslides and debris flow."[1]

The Los Angeles Times *bicycle club on Western north of Pico Blvd., 19895. Photo courtesy of the USC Libraries - California Historical Society Collection*

Living here for twenty-seven years, the Stacey family love their river rock home. I asked Mrs. Stacey if they had earthquake damages. "No earthquake problems, we just moved in in 1993 and were greeted by the 1994 Northridge quake. No damage, no cracks, no falling rocks. It shook, but the home is solid," she said.

Around 1908 the *Los Angeles Times* reported that "The movie makers come at first to escape snow and ice... But the bright quality of the sunshine and the number of clear days in which they may work, together with variety of scenery, has all been found ideal."[2]

By 1917 the manufacture of motion pictures had become the United States' fifth largest industry. The vast majority of the films were made in Los Angeles though movie filming began in New York. There, unions, permits, and especially the weather restricted filming throughout the year. "The overriding reason for the migration of the movie maker" from New York to Hollywood, John and La Rae Caughey wrote in in their 1977 history of Los Angeles, "was the same reason that attracted normal Americans," (while some may question the movie industry's normality)– "sunshine...cameramen had to rely on the sun for in the early years intense stage lights had not been invented."[3]

Later in the 1920s, individual film-crew members

Aleesie Mallette residence 1929, (no address noted and may be demolished) photo from San Fernando Valley History at digital-library.csun.edu

4146 Klump Street, North Hollywood. The Bowman family home was completed in 1927.

Family photos sent to Albert Knight, archeologist, by Mrs. Bowman.

would buy a lot and settle down in their new land of dreams to build their home. A few used their creative skills and built imaginative homes in stone from the nearby L.A. River—"the only river I could walk across," Samuel Clemens boasted.

Individualists moved to SoCal and Bernard Bowman Sr. showed his independent thinking in building this beautiful home. A grading contractor building new residential lots and dirt roads, he saw a river rock home during his work, possibly at Stonehurst. The grading equipment used to clear lands of chaparral and prepare dirt roads consisted of two mules pulling a wide metal blade.

Bowman hired a mason to build a river rock home for his young family a hundreds yards from the Los Angeles River. "Grandpa wanted a similar rock house," granddaughter Pam Bowman wrote in her letter.[4] His story gives us a rare glimpse of the builders of these stone homes as few records of these people remain.

In the late 1990s Bowman's 86-year-old son visited the home and reminisced how he and his next door neighbor used to "go skinny dipping" in the Los Angeles River, just south of the home, when people could enjoy the river with its flowing water. Today, people drive in the concrete river.

In the late 1920s "Not many houses around then, lots of orange groves, peach trees covered most of the land in the neighborhood along with unsold lots and a few wood framed houses." The granddaughter remembers her dad "speaking fondly about the rock house", an excellent and rare example of an arroyo stone residence in Studio City. "Buildings constructed of arroyo stone are extremely rare in this part of the city," she wrote. The Bowman family home was completed in 1927, but "unfortunately," Pam wrote, "they only got to live there about three years." Depression struck. In the Depression years, the family needed to move and they landed in the Mojave Desert area, and her grandfather, the grading contractor, began mining for gold.[5] Mr. Bowman died while living in Victorville, he was 82.

11213 TIARA STREET, NORTH HOLLYWOOD

A modest 850-square-foot home built in 1923 remains vulnerable to demolition as it is caught up in the zoning for apartments along with the designation of the

area as a state enterprise zone, one of only three areas in Los Angeles that gives tax incentives to businesses. Development of this property would require the demolition of a significant local historic building, though this stone home is not registered as 'historic'.

I could not see where the mason found the stones as this location is remote from washes and located in the plains of San Fernando Valley, where the dirt does not have many stones. The base of the rock walls were built with that smooth taper popular in craftsman homes and indicating thick stone walls. To save on the cost and

539 Harps Street, City of San Fernando

effort of building wood forms for the home's windows the builder made the two front windows are the same size.

THE CITY OF SAN FERNANDO

Far off from the hustle of Los Angeles lies this small city on the northeast corner of San Fernando Valley. In 1913 amidst strong controversy which lasts to this day, Los Angeles City brought water from the Owens Valley on the east side of the Sierras via a 233-mile aqueduct to Southern California. Many new rural towns depended on artisan ground water wells then and these wells became either dry or did not supply adequate water to handle the town's growth.

In its largess, the movers and shakers of Los Angeles City at that time "allowed that formerly independent city to use its water but at a price."[6] It had to be incorporated into Los Angeles City. San Fernando City, the home of the San Fernando Mission, said no thanks, we have our own water, which they obtained from their own artisan wells, fed from the slopes of the San Gabriel Mountains.

STONEHURST

"The new foothill community where life is worth living."

So bragged Pep Rempp's promotional pamphlet on his new development.

They were built in a unique era, the river rock buildings, and tell us of master mason craftsmanship, when mastery was admired and appreciated. These buildings also hint to their pride and their dreams. "We don't know enough about our history here in Southern California," Huell Hauser reminds us.[7] And that is true of the builders of these stone homes and the rare documented story provides glimpse of life during the early twentieth century,

An historic photo, taken looking south across the San Fernando Valley, of the original Stonehurst Tract. Security Pacific archives. Three rows of identical stone homes with stone garages on their deep lots in the edge of undeveloped San Fernando Valley.

so different from our experiences. And yet one condition remains constant. Earthquakes. Over the past century seven events did little damage to the river rock homes while inflicting catastrophic destruction to freeways, wood-framed homes and apartment buildings.

With the other stone homes we've seen, the common practice would be to buy your lot, whether a half-acre or smaller, and then build your home either yourself or hire a mason. You decided on its shape, size, and appearance; but Stonehurst tract, built in 1923-1924, stands out as the only group of identical prebuilt stone homes. Sixty small homes were built side by side on deep lots to be sold to the new arrivals of eastern migrants.

Stonehurst was built in the 'sticks', far from downtown Los Angeles and other towns.

In the early 1920s, one saw barren land with dried native grasses, scrub brush, coyotes, snakes and "... small boulders and rocks of all sizes," archeologist Albert Knight writes. "When the right moment arrived, it would be this readily available native stone material-most likely light colored granitic 'found river rock' - that would provide the building material for the Stonehurst neighborhood."[8]

Many rural areas of Southern California were accessible by 'the red car', the new privately-owned electric trolley system that transported people and goods throughout Southern California. And by 1910 this narrow gauge rail trolley system ran on a thousand miles of track. Henry Huntington developed the red car transportation network and it extended from Riverside to Los Angeles, from the beach towns of Santa Monica, Manhattan, Hermosa, and Redondo to Los Angeles, from Long Beach to Van Nuys in the south-central side of San Fernando Valley. It transported mail, sugar beets, oranges, tourists, and commuters who wanted "to sleep in the woods and have an office in the city."[9] Stonehurst became one of those places. Actually in the 'sticks' on the north side of the Valley, transportation to this new tract occurred by car as the red car stopped in Van Nuys, some eight miles south.

Stonehurst is probably the first tract housing in California, but certainly the only tract of stone homes; it remains a treasure built with skill, artistry, and talent. Due largely to the efforts of Dr. Albert Knight, anthropologist and former owner of a Stonehurst home, this unique tract was included in the Los Angeles City Historical Monuments directory.

The stone home prices varied, for the buyer had choices: Did he want to have a Model T Ford in the garage? Or would he like his new home furnished? "Arranged and

The first tract of homes in L.A. and they are made of river rocks. In this photo the homes need windows, doors, walkways, and the street.

In the photograph above, the tract is completed and ready for new owners to settle in. (From L. A. Library public photo collection.)

set-up for the mid-west immigrant who moved to SoCal with only a suitcase."[10]

These historic photos show rows of identical modest homes completed before the necessary access roads and walks. Forty homes were built along with the stone community center in Phase I, while Phase II completed the neighborhood. All the homes had deep lots, a half acre, the size advertised that allows the family to grow and farm adequate food to support a family of four. A few families did grow crops of melons, corn, and beans and raised chickens, pigs, and horses. A favorite pastime for young boys was to learn firearms and improve their marksmanship by shooting rabbits and rattlesnakes.

9901 DRONFIELD, LOS ANGELES

Possibly because of its remoteness, Rempp in 1930 built a stone community center at Stonehurst Park. There the residents enjoyed BBQs, family picnics, square dancing, and baked pies. Other foods served included goat, chicken, and calf meat, and bread cooked in a wood fired stove. "The Stonehurst Recreation Center," Floyd Bariscale wrote in his blog, "came pretty close to being demolished in 1977. The City had planned to raze the building that fall after opening a brand-new 10,000 square-foot community center." Local residents and the district's council member balked at this and it became declared a city landmark in spring of 1977. [11] (The 1970s was a dangerous decade

Stonehurst Community Center, built in 1930. L.A. City finally recognized this building, responding to local citizens demands, as Historic Cultural Monument #172 in 2008.

A typical Stonehurst home built in 1924 and photographed in 2015. This building has remained true to its original construction with the front porch enclosure being the only noticeable change.

for the historical buildings of Los Angeles. Many were demolished for bland apartments and strip retail stores.)

All the buildings were built of river rocks: the small post office which connected the remote tract to the wider world, the country store which sold necessities, the sales office selling the homes, and later the community center. Although other building materials were available then, such as wood, adobe, and brick, Dan Montelongo, the mason, and Pep Rempp, the developer, chose cobblestones.[12] A former resident, Mr. Hernandez, mentioned in an interview that the inside of the stone cottage was finished with wood-framed walls, wood lathe, and smooth plaster, possibly to appeal to the feminine tastes.

"A wide variety of rock types are found in the (Tujunga Canyon) washes," Knight wrote. "These are mainly metamorphic (granite, gneiss, quartz)," and the materials found across the entire area of the large and millennia-old alluvial fan below Big Tujunga and Little Tujunga Canyons "include countless millions tons of rock, gravel, sand and large boulders."[13]

Centuries ago the Tujungans, meaning "Old Woman Place", occupied this land at the base of the Tujunga canyons. "In 1992, a (Stonehurst) neighborhood boy found an arrowhead in his backyard at a depth of three feet," Knight wrote. He estimated that the arrowhead could be 2,000 to 3,000 years old.[14]

The prolific Mescalero Apache mason Daniel Montelongo would ultimately build over a hundred stone structures using minimal help, taking his Model T pickup into Tujunga Canyon to select and then load the stones, return to the job sites, and unload the stones. A local grader also scraped up many stones and dumped them near Montelongo's work. Rempp advertised and sold the finished homes for $1,800 unfurnished and $3,200 completely furnished and outfitted with kitchen ware along with a Model T in the garage. The intent was to make it easy and convenient for the newcomer to settle in without shopping, just unpack their bags and cook dinner.[15]

The promotional pamphlet advertised that "three adjoining tracts will be opened this year (the three rows of stone homes are shown in the historical photo on page 256 of this chapter). A different style of residence will

The one house in Stonehurst with a mission tile roof, it sold for $4,500 furnished including the Model T in the garage. Mr. Rempp's secretary was the first resident.

Right: ***The post office building is now a residence. Built in 1923 at 10103 Stonehurst Avenue; this is one of two buildings in the community built with arches.***

Above: ***A small but typical stone garage sized for the model T Ford.***

Right: ***Close-placed interlocking stonework shows the handiwork of an artist mason who selected his cobblestones intentionally. Notice the variety of patterns in the stones, placed like a painter choosing just the right colors.***

This the largest home in the Stonehurst community. It illustrates the fine art of river rock poetry.

grace each street. Each residence on Sheldon Avenue contains a fireplace and built-in features, is furnished to suit the purchaser, and all are of the same size and architecture (as if that was a desired feature) In addition to the house being completely furnished, (This was done by the developer Rempp with awareness that his buyers were coming from the mideast by train having left their homes and furnishings.) a Ford car and a garage are given to each purchaser, though buyers did have the option to buy an unfurnished stone residence and an empty garage. The typical home had 625 square feet of floor space on a large lot, 60 feet wide and 330 feet deep. The home's stone walls taper from 36 inches thick at the base to 12 inches at the eaves.

Later Rempp was arrested for embezzlement; he kept most of the money from the home buyers and failed to pay his creditors. "Until one day sheriff's deputies arrested him and threw him in jail, never to be heard

Dapper Dan Montelongo as a young man, this photo was probably taken at his wedding.

A young Daniel Montelongo before WWI in his vaquero outfit. Photo provided by his daughter and found in the Albert Knight Stonehurst research at CSUN.

from again," said an article in *The Daily News*. "It took ten years to clear title to each home,"[16] an unfitting end to this unique project.

Dan Montelongo built all the homes and garages. At times a couple helpers joined him; on later stone buildings his ten-year-old son helped, usually mixing the Portland Cement mortar. Montelongo's masonry skills shine in this stone tract adding artistic details to both the inside and outside of the residence. Huge stone slabs would decorate the hearths, while outside sunbursts of stones appeared randomly on building and garden walls. Then, as seen at the post office building, refined well-proportioned arches constructed using thoughtfully selected thin stones. "Many considered Dan to be an artist," Knight said.[17]

In the next chapter we'll meet another artist working in stone, George Harris in La Cresenta. As the two masons built their stone structures during the same time it is fascinating to speculate that Montelongo of Stonehurst fame and George Harris of Cresenta Valley fame may have been friends, both being master artist-masons, and they may have assisted each other in building these excellent river rock structures.

10021 Stonehurst Avenue

Montelongo "may have been the most prolific of all the local stone masons," Knight writes, and that "he could have built over one hundred stone buildings in his life (1895-1944)...The fundamental difference in the nature of the stone tells a little difference between the two master masons. The fieldstone used by Harris was only a little worn," for he gathered his unweathered stones from the hillsides, "closer to the mountains, and the stone in his structures is usually set with a fairly flat face (Bolton Hall for example which we'll visit in the next chapter.) "The

Modest river rock home at 11213 Tiara Street, North Hollywood.

The sunburst circular narrow stone design appears in many homes and may suggest a intergenerational guild sort of relationship between the many masons in Southern California.

river rock used by Montelongo is usually rough stacked, and usually has a relatively uneven (wall) face."[18]

While reviewing Knight's work at CSUN's Oviatt Library, we met Julie, its archivist. She coincidentally lived in a stone home in La Cresenta and shared her family's experience of earlier residents. In the early weeks after her family moved into their new river rock home, her startled daughter asked, "Who's that man that wanders by my bedroom window?"

Reasonably Julie thought, stalker, and went looking

Although in the Stonehurst Tract at 10021 Stonehurst Place, this home must have been especially commissioned as it is larger with arches and and an entry porch

10107 Stonehurst Place built with attention to detail as seen around the large arched window and the tight placement of the stones.

for that man, finding no trace, no footprints in the dirt, no snapped twigs, nothing. A few days later a neighbor with a house warming gift came by to welcome them to the neighborhood,

"How do you like it here?"

"Well, it's very pleasant and quiet, my daughter has made new friends at school but..." She paused to frame what she said next, "There seems to be a man wandering outside her bedroom window occasionally. We try to find him, but he disappears."

"Oh, don't worry, dear" the neighbor replied, "that's our old postman, an earlier resident here. Though he died a while back, he occasionally wanders by. Could be he's checking that everything is okay."

From A. Knight research: "It reportedly served as a speak-easy (a place where liquor was served), until a few years later, the story relates, "one customer was shot by his wife 'who caught him in the act.'" America became obsessed with alcohol during the early nineteenth century. While busy orating its evils and legislating it use, Americans used more of it, made more of it. 'Speak easy' bars proliferated like weeks and many a cobblestone building turned into a secret bar during Prohibition years. Earlier years were tough on people's spirit for their suffering through WWI deaths and sacrifices, through a recession, and through a pandemic. No wonder the 'Twenties roared.

10004 CLYBOURN AVENUE, SHADOW HILLS

Built on a knoll in 1913, this was one of the earliest stone homes in the area and built on the flatter lands. But this one is perched on a hilltop overlooking the valley. Its design includes a neoclassical styled veranda and the stones laid up flat faced suggests this is a home of veneered rock construction and quarried stones, which requires shaping the granite stones. Both these quarried stones and the hilltop location suggest wealth.

A brother of William Mulholland of Owens Valley aqueduct fame, Stan Mulholland lived here enjoying both the expansive view and a less hectic life than William, who, with thousands of workers also completed its construction 1913.[19] These years saw substantial industrial growth in the U. S. and another massive engineering project involving water celebrated its completion in 1915 with the inaugural ceremonies for the Panama Canal.

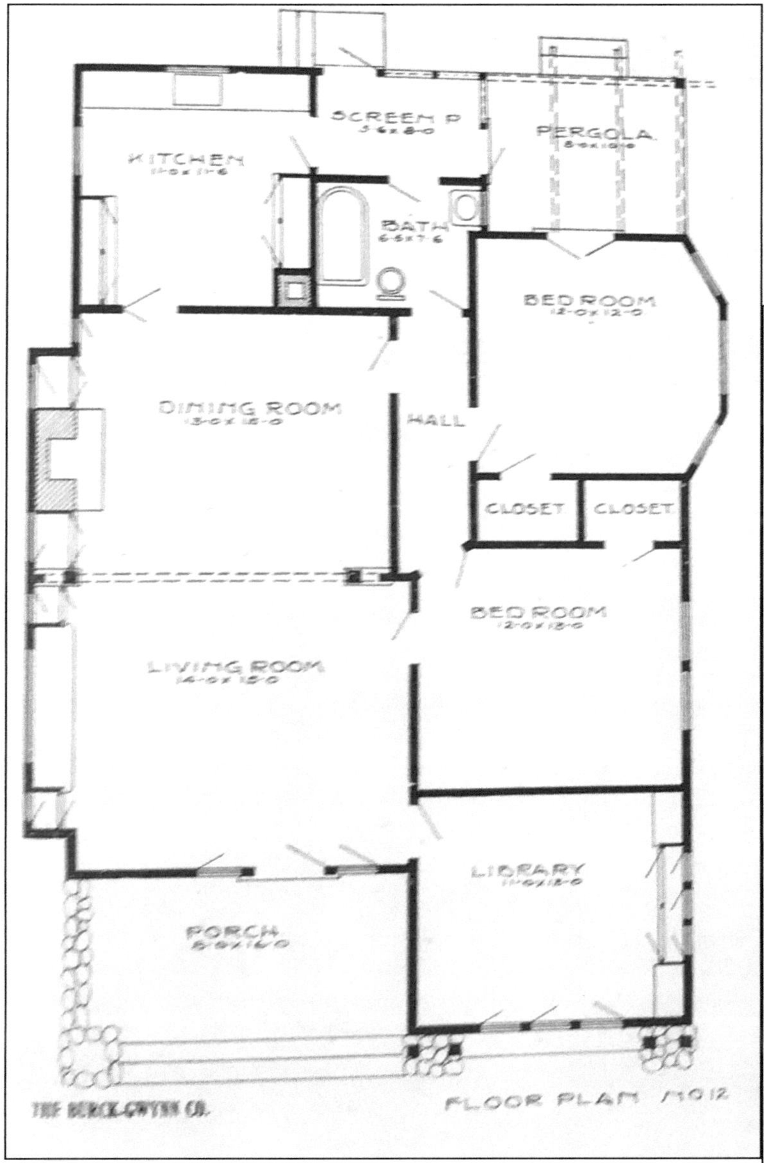

The typical modest cobblestone home

No. 12.

The artistic use of cobblestone is the feature of this house. The wide cement steps lead up to a sheltered porch of generous size. Entering the front door, we find a long room, divided only by columns and a low beam. The long square bay window on the outer wall is taken up with book-cases and a hinged seat in the living room, with low china cabinets and a tile mantel in the dining room.

The pergola on the rear makes an ideal outdoor sitting room, and is convenient to the kitchen, should it seem desirable to serve meals there in pleasant weather. There are French windows from the rear bed room.

The library has a book-case with leaded glass doors, under a group of small casement windows, with a seat on each side. The room is well lighted by a second group of longer windows on the front. The main rooms have oak floors.

The door from the bath to the screen porch does away with the necessity for any extra plumbing and lightens the expense.

Complete plans, details and specifications will be furnished for $11.00.

Los Angeles Historic-Cultural Monument No. 135 is a former stable built in the Mission Revival style in 1936 by actor Francis Lederer. You really need to know as authentic as the building may appear, it's only about seventy years old. It does not date to the Spanish colonial days.

When walking down the central corridor you might begin to visualize the building's history as horse stables, with the central dirt aisle covered with hay and the horse stalls on both sides.

LEDERER STABLES, 23130 SHERMAN WAY, WEST HILLS, SAN FERNANDO VALLEY

Originally this river rock building housed the Lederer horse stables, then it became a motion picture school, the Canoga Mission Gallery and Community Art Center, and today is occupied by the Malibu Wine Company. Francis Lederer became a famous actor in the 1930s and built his stone home a knoll overlooking the stables. The home is now gated, remote on its hilltop from the commercial activity along Sherman Way.

A stone home I found just driving through the town sits quietly on Fillmore Street. The mason carved the address '13333' in the large boulder by the front gate entrance.

13333 FILLMORE, PACOIMA

This stone work is fun and delightful. Note the standing stones on the top of the wall above a circle of thin oval stones set on edge outlining the attic vent. And just below that vent a line of thin stones form the top of the window with a tall thin boulder for the keystone. This is a prime example of the river stone mason's art yet Los Angeles city does not list this beauty as a Historical Cultural Monument nor would a historic preservation review be required of a developer intending to demolish this 1,390 square-foot stone home.

Stone flowers adorn the home above the windows and the doors.

L.A. *Historic Cultural Monument #644: "The Stone House"*
Built in 1925-26 by local artisan Dan Montelongo, of Stonehurst fame, this house contains 1,345 square feet with two bedrooms and one bath.

8642 SUNLAND BOULEVARD, SUN VALLEY, SUN VALLEY YOUTH ARTS CENTER, LOS ANGELES CULTURAL AFFAIRS DEPARTMENT

Then and now photos: the photo above taken from the website flickr probably taken before 1998 when the building was sold (notice the graffiti on the roof) and below from the city website on this building taken after renovation by Los Angeles City for the youth arts center.

11306 LAUREL CANYON BOULEVARD, LOS ANGELES

Based on the fine masonry work, Dan Montelongo may have built this home in 1930–1932. The 2,750 square-foot size and four bedrooms with two bathrooms home indicates this home was built for a well-off family. Again Knight judges this as "Easily the best surviving Montelongo structure."

In the city records I found a historic building permit for the two car-garage that cost $300 to build. The owner signed his name to the permit as R. Keith. Since then, the Franciscan Sisters bought it in 1968 for their retirement home and their thoughtful maintenance and additions have kept it "A rare example of an arroyo stone house in Mission Hills," praised Historic Places of L.A.

END NOTES

1 Deverell, W. and G. Hise, *Land of Sunshine*, University of Pittsburgh Press, 2005, p. 39.

2 Weaver, J., *Los Angeles the Erroneous Village*, Capra

Press, 1980, p. 80.

3 Caughey, J. and L.R. Caughey, *Los Angeles: Biography of a City*, University of California Press, 1977, p. 224.

4 Bowman, P., Letter to Albert Knight in 1997, California State University at Northridge, Oviatt Library Special Collections.

5 Ibid.

6 Caughey, J. and L.R. Caughey, *Los Angeles: Biography of a City*, University of California Press, 1977, p. 276.

7 Hauser, H., "California Gold," PBS SoCal, February 9, 2019.

8 Knight, A., "Stonehurst Historical Survey, Draft," April

At the mouth of Tujunga Canyon wash, an early settler built this stone home that is now part of the Lakeview Terrace Recreation Center and Equestrian Park. Built by the Army Corps of Engineers this park is located on the north side of Foothill Boulevard.

9810 Wentworth, built as a smoke house probably in the 1910s. The large chimneys of stone have survived a few strong earthquakes suggesting that their size may have been the reason for surviving. Photo taken by Al Knight in 2004

2007, California State University at Northridge Special Collections.

9. Caughey, J. and L.R. Caughey, *Los Angeles: Biography of a City*, University of California Press, 1977, p. 221.

10 Gebhard, D. and R. Winter, *An Architectural Guidebook to Los Angeles*, Angel City Press, 1965.

11 Bariscale, F. "Blog #172," Big Orange Landmarks, ww.bigorangelandmarks.com, August, 13, 2008.

12 Knight, A., "Stonehurst Historical Survey, Draft," April 2007, California State University at Northridge Special Collections.

13 Caughey, J. and L.R. Caughey, *Los Angeles: Biography of a City*, University of California Press, 1977, p. 247.

14 Knight, A., "Stonehurst Historical Survey, Draft," April 2007, California State University at Northridge Special Collections, p. 33.

15 Sesma, C., "Historic Stonehurst District," The Valley Journal of the San Fernando Valley Historical Society, vol. 67:9, September 2017.

16 Timmerman, J., *The Daily News*, November 26, 1981.

17 Knight, A., "Stonehurst Historical Survey, Draft," April 2007, California State University at Northridge Special Collections, p. 33.

18 Sesma, C., "Historic Stonehurst District," *The Valley Journal of the San Fernando Valley Historical Society*, vol. 67:9, September 2017.

19 Knight, A., "Stonehurst Historical Survey, Draft," April 2007, California State University at Northridge Special Collections, p. 33.

CHAPTER 16

Cresenta Valley

Home to the communities of La Cresenta, Montrose, Sunland, Tujunga

La Cresenta Valley looking east, the San Gabriel Mountains are on the left and the Verdugo Hills on the right. This aerial photo taken in the 1930s was found in the Spence Air Photo Collection at UCLA Dept. of Geography.

Geologically Cresenta Valley is unique. Squeezed between the San Gabriel Mountains on the north and the younger Verdugo Mountains on the south, this crescent-shaped valley extends east fourteen miles from the San Fernando Valley to drain into the Arroyo Seco at Pasadena.

This long thin valley, only one and a half to three miles in width, became in the early twentieth century home to the largest collection of river rock structures. Torrential rain storms filled this space between the mountains. Canyon muds, cobblestones, and debris, unable to flow further to the south, became a compressed alluvial fan.

On a nice Sunday afternoon in 1912, Los Angeles residents might take their new Model Ts out for a drive in the country. The Automobile Club published several pocket size maps to guide them on a scenic route. This map guided drivers from downtown Los Angeles north to Sunland and back along a dirt road on the east side of the Verdugo hills. At that time, the drive of fifty miles might have taken three hours. Pocket highway maps aided less leisurely road travel than the strip maps of 1912. Hi Wa "Strip Graphs" were produced in 1947 to guide drivers on modern highways to cities hundreds of miles away. Library of Congress online exhibition: www.loc.gov/exhibits/lamapped

When the storms became powerful enough and long enough this river of alluvial mud would drain eastward to the Arroyo Seco, carving the canyon deeper and deeper as it flowed south into the Los Angeles River.

Young, small, and still growing, the Verdugo Hills trend east-west just like their older and much larger sibling, the San Gabriel Mountain range. These Verdugo Hills grow from the squeezing of the Sierra Madre fault on its north and the Verdugo fault on its south. Earthquakes in either of the faults uplift the block of granite that forms its core. As such, its rocks are similar to the granite igneous rocks of the San Gabriel Mountains.

Mexico granted Jose Maria Verdugo this land in the mid-1700s to become the Rancho La Canada, 'the Valley' in Spanish. I believe that due to this unique geologic arrangement of being sandwiched between two parallel mountain ranges that provide unlimited cobblestones, dozens of river-rock homes were built here, making Crescenta Valley a supreme home for this indigenous architecture.

This valley became the new home to many settlers from Virginia, Missouri, Idaho, and neighboring states where the traveler, in the late 1800s, rode on a steamboat

Early Tujunga being settled on the alluvial fan created by mud flows from the Tujunga canyon.

up the Mississippi River to the Missouri River and then caught the new Union Pacific transcontinental train to ride in an unheated wood pullman car. They sat and sleep on the same wooden bench for three days, selected volunteers to watch the cast iron stove during the cold nights while traveling through the Sierra Nevada Mountains towards San Francisco. This tumultuous city on the hills was not their final destination, for they wanted to live in the land of sunshine, not the hills of fog.

They would stay a few days in wood-framed boarding houses waiting for the weekly ship to Los Angeles. After a week of sailing down the Pacific Coast, they arrived in San Pedro Harbor, or at one of the many piers along the Southern California shores such as in Redondo, Manhattan, and Santa Monica. From there they traveled by wagon in the late 1800s to the busy town of Los Angeles and stayed in a rooming house or hotel. Then a land promoter took the settlers by a six-passenger wagon decorated with a fringed canvas cover with bright banners and pulled by two horses. They arrived at the land sale and the promoters' gala celebration of barbecue, band,

In this early photo of the Tujunga area looking north towards the Tujunga canyon in the San Gabriel Mountains. Seen here is the sloping alluvial fan spreading left and right. The streets are dirt and Bolton Hall lies alone in the center of the photo.

and more banners.

La Cresenta Valley held many of these land sales, and its 2,000-foot altitude appealed to both the retired farmers for its remoteness from the noise of Los Angeles, to the 'health seekers' for the dry air, and to all for the water from the canyon streams and the ground aquifers. Here rock homes were built, as well as wooden homes, many which did not survive the foothill storms. Communities were settled with stores, churches, and schools. The earliest schoolhouse was built with wood around 1903 to handle the first class of six children taught by a young woman.

A common sight during the disastrous flood of 1934

In the early 1900s, hardworking people created the foothill communities of Sunland, La Cresenta, Tujunga, Montrose, La Canada, and Flintridge. Using the rounded rocks that had tumbled down from the San Gabriels over millennia to clog this land like no other area in Southern California and binding them with the recent invention of Portland Cement, the settlers labored to build their new homes. Being physically demanding work, those folks slept well, lived long, many into their 80s and 90s with little to no medical services.

Today we may walk or drive on on their land unaware of their culture of neighborly friendship and devotion to religious standards. Yet the settlers were unaware of the unique forces of this new land, so different from the flat lands of the midwest. They were soon to be challenged by fires, floods, drought, and earthquakes. Newspapers and a few original letters showed that they lived simple uncomplicated lives, but difficult. Either the vegetables

This stunning river rock home represents many of the valley's stone homes of the 1920s. The rich details around the windows show the mason's artistic ability.

One Of The Original Little Landers

MISS MABLE HATCH relaxes in front of home she and her father, Hiram F. Hatch, a Civil War veteran, built shortly after coming to Tujunga in 1913. Mable, aunt of Sunland's Ashley Hatch, never married and Ashley made his home with her for years. Home was located on site of present Tujunga Post Office. For over 30 years the sole caretaker, custodian and manager of old Tujunga cemetery donated by Tujunga's developer, M. V. Hartranft, Miss Hatch was loved by all, operated her own insurance business in a small office next to old Valley National Bank at Commerce and Tujunga Canyon Blvd. (Building is now vacant.) She told of her first view of Tujunga after a long, hot buggy ride from the train at Roscoe Station (now Sun Valley). She said she and her father stood on the rocky, treeless lot they had purchased and cried. "It was so baren, so unfriendly and so unlike the rolling, green hills of our native Michigan. We could hardly bear it in the beginning. But we made the most of it and soon came to love the west and the hilly slopes of Tujunga. We made our home here and never regretted it." Miss Hatch died in August, 1957, at the age of 77.

grew or they didn't. Either the home was watertight or buckets were brought in.

Interested and invested in their new community's well-being, the settlers attended community meetings, of which some handwritten minutes have been saved. Absent in those minutes are issues of crime or conflicts. Present are simple notes like, "After his presentation on the 1st National Park (Yosemite) we went home." Homes had minimal electricity, minimal closet space, a wood burning cast iron stove for cooking, and a rock fireplace for heating. When sun set so did the homeowners. When the sun rose the day began. Simple.

Regina, of the Little Landers Historical Society, told us that "these were people of performance, not much

mulling around. These were just-do-it people; community builders who began new lives organizing and socializing through clubs, church activities, dances, and barbecues." A founder of La Cresenta and member of the Little Landers colony of the 1900s, Mabel Hatch wrote with a succinct style in capital letter worth presenting in full:

> *"Every week we went to town marketing. It was held in the club house and we all lighted our kerosene lanterns and thudhed (sic) up the hill. We were weary in mind and body, running behind a plow and throwing out rocks only to find when you go back over the field to plant your walnut trees you have to dig the holes with a pick, is neither refreshing nor encouraging. Furthermore, you take so much stone out of the hole, there isn't enough earth to plant the tree. But we went to the town meeting."*[1]

An interested viewer today could not visit all these gems in one day; we've located eighty stone homes in the Cresenta Valley communities, by far the greatest concentration in California and seventy-five of them were built in the decade of the 1920s and the early 1930s.

October 29, 1915. Garden meeting at Bolton Hall. Photograph by J, Lamson. Donated to the Little Landers Historical Society by Mrs. Lee Kratbe in October 1994.

GEORGE HARRIS

Mason's hands. Dry, cracked, strong thick fingers. And calloused. George Harris had those hands; he was just such an alchemist. Using unmodified cobblestones of granite, he constructed structures in Southern California of subtle texture and color, stone buildings that are as fresh today as when built a hundred years ago. Harris' cobblestone homes and churches are not random piles of granite stones magically standing for a century. He possessed a stone mason's skill, learned from years of assembling river rock structures, to create the strength, form, and stone's arrangement needed for durable construction.

Progress would have been steady. Two or three masons working together building a section of wall where one gathered, cleaned, and sorted the stones, another mixed the mortar, making a mix of three parts sand to one part Portland cement; dryer in the wetter winter months and wetter in the dry summer days, and of just the right consistency that the mortar withstood the weight of the boulders. A master mason with keen perception selected the irregular shaped stones so they would fit together tight to minimize the use of the expensive Portland cement. In this way a twenty-foot stretch of wall three-feet high and eighteen-to-twenty-four-inches thick could be built in a day; lifting and placing about fifteen thousand pounds of solid granite plus the weight of the mortar. All the cobblestones remaining in their found

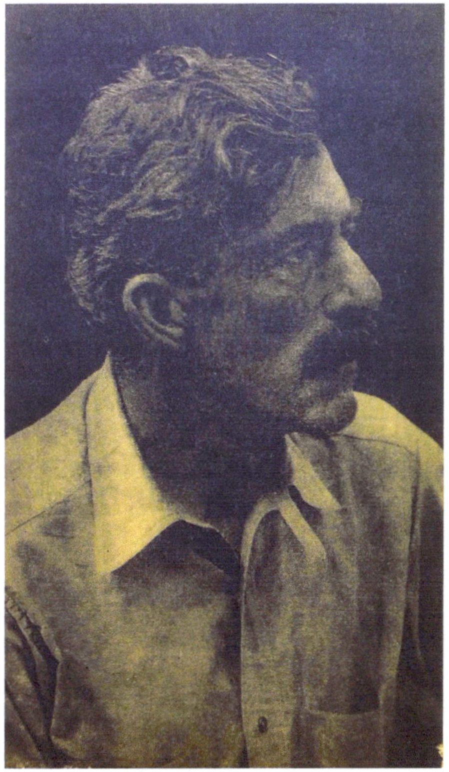

George Harris and his masonry tools.

This stone home ascribed to be a George Harris home. Below; A George Harris magazine advertisement.

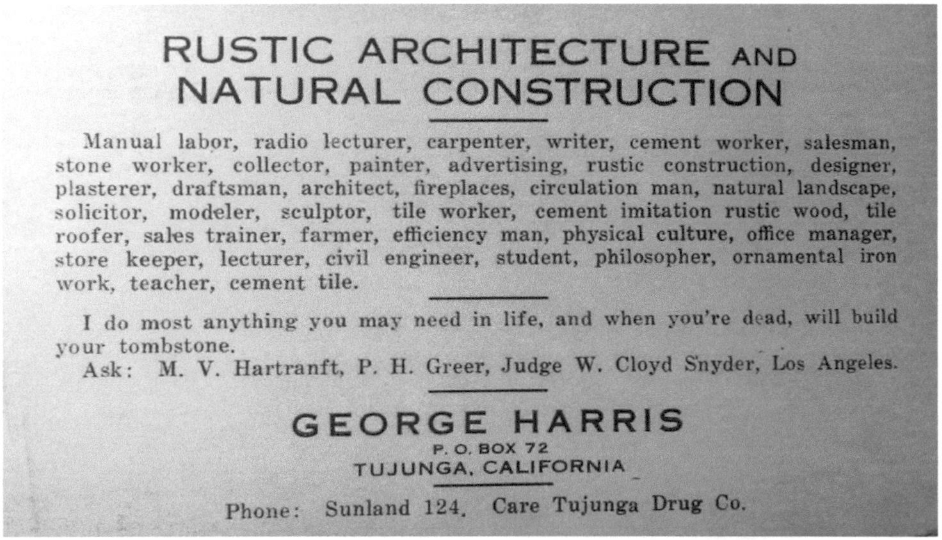

RUSTIC ARCHITECTURE AND NATURAL CONSTRUCTION

Manual labor, radio lecturer, carpenter, writer, cement worker, salesman, stone worker, collector, painter, advertising, rustic construction, designer, plasterer, draftsman, architect, fireplaces, circulation man, natural landscape, solicitor, modeler, sculptor, tile worker, cement imitation rustic wood, tile roofer, sales trainer, farmer, efficiency man, physical culture, office manager, store keeper, lecturer, civil engineer, student, philosopher, ornamental iron work, teacher, cement tile.

I do most anything you may need in life, and when you're dead, will build your tombstone.

Ask: M. V. Hartranft, P. H. Greer, Judge W. Cloyd Snyder, Los Angeles.

GEORGE HARRIS
P. O. BOX 72
TUJUNGA, CALIFORNIA

Phone: Sunland 124. Care Tujunga Drug Co.

natural rounded shape, neither chiseled nor quarried, the completed wall came alive. These solid interlaced granite stones began to move with their waving patterns and freckled faces reminding the onlooker that these granite stones once where liquid, millions of years ago.

And decades ago, five masons cleaned up and, dressed up in their fine clothes, posed for a photograph placing the cornerstone for Bolton Hall community center in La Cresenta—a traditional ceremony in 1913. Nine months later the building was completed. George Harris, the master mason, wore a clean white shirt, clean jeans tucked into scuffed shin length leather boots, and posed casually resting his hand on the cornerstone, proud of this new community building. He should have been proud, for he designed this natural looking rugged stone and boulder structure, and after seven months of intensive labor these men completed Bolton Hall.

Connecting so closely with nature in his work, a stone mason's character tends to be laconic, quiet, thoughtful, allowing him to focus on selecting the right stones. And this, I believe, is not taught, but inborn. Sitting in the patio, I imagined George Harris working here, building the center in the new Little Landers community, guiding his assistants to locate particular stones, telling his mortar maker to use less water or to add water.

In Albert Knight's unpublished work on Stonehurst he wrote of Harris's building philosophy. "'My art consists in following the line of least resistance...the natural path (an animal path) is a crooked one and not a combination of geometric curves and angles.'"[2] Building with the cobblestones "'is an evolution...no practical working plans can be made in advance...no two pieces (buildings) are alike.'" He continues, "' Bolton Hall was built almost entirely without a plan.'"

George Washington Harris's buildings, furniture, life and philosophy is vast enough to fill a biography, one that has been waiting the hands of a gifted historian; yet one has not been written.

William Symthe and Bolton Hall founded the Little Landers commune around 1913, a time of social unrest when in the U. S. idealized self-sustaining communities, or communes, of a few hundred people became popular.

In an earlier chapter we met Lee Pitzer who settled in Claremont forty miles east of La Cresenta. Both Pitzer and Harris created communities, both listened to the stones' music and created buildings that remain, refurbished,-occupied, and loved.

Laying the cornerstone of Bolton Hall, April, 1913

Steinbeck gives another possible reason for building with cobblestone and boulders, "To the people of the valley the slate roof (think 'rock home') was the show piece of the county...This man was steady." And indicated he was staying for a long time. The more restless people and those looking for a quick buck who were unable to stay in one place, built flimsy houses of wood siding, single thickness drafty, dusty, and temporary.

INTERVIEW WITH BOLTON HALL DOCENTS SHIRLEY AND JUDY

Shirley and Judy, docents at Bolton Hall, the name of the community building and the visionary of 'the Little Landers' commune, described the founding of the Little Landers community: "Mr Hartranft was a developer so he bought this big chunk of property," Sheila said, "Many acres, don't know how many, but this whole area of Sunland/Tujunga were two little towns. This one (Tujunga) was started as a communal town where as long as there was water settlers could irrigate their crops and themselves and live here. That idea was from Mr. Bolton Hall. His plan was to make a community of people that would barter." The settlers arriving "from back east were told that with a shovel and bag of cement one could build themself a house.

"Settlers came by the thousands." she continued "arriving by train for railroads had been completed at this time." "1910", Judy added. Shirley continued to describe the settlers, "People were looking for land. Something they could settle on, something special. So when (they)

Bolton Hall, 10110 Commerce Street, originally this stone masterpiece was named after Mr. Bolton Hall, a moving force behind the Little Landers communities, and the name became fitting for a community hall. Later it was occupied as a city hall with a jail and is now owned by the City of Los Angeles with its long-time occupant being the Little Landers Historical Society and its museum. A fun and educational visit to Bolton Hall is not to be missed. George Harris penned a letter after he had finished building Bolton Hall, which tells us of his sentiments towards his great creation. "This house is what it is because of its setting. (echoes of Frank Lloyd Wright). It is constructed...of natural materials; the stones gathered from the hillside and selected and placed in natural positions." His assistants in building Bolton Hall were "three young men from the Balkan countries." Marshall V. Hartranft provided the financial support for the building. [3]

The corners required unique stones, river rock with two sides at right angles, natural not chiseled. And notice the arrangement of the stones of varied colors, shapes, and sizes.

1914. Interior of "The Clubhouse" (Bolton Hall). Note the copper plaque on the wall and a rope for the bell.

heard about this place, Marshall Hartranft advertised in papers and whatever, they came here...Three of them were a team, with George Harris as the builder, Hartranft was the developer and Mr. Smythe," who started the community," being inspired by Mr. Bolton Hall, the writer. Bolton helped start the community but he didn't stay and this town of Monte Vista lasted about a year.

"In the meantime this building (Bolton Hall) was built in seven months (some sources say nine months in 1913). It has been used for dancing, for a church, and a police department after they put the addition onto it (on the back side beyond the large boulder fireplace). It was also the health department." Judy, a lifetime resident of Tujunga, remembers getting her polio shot here.

"It was used for musical events, singing," she tells us; and in those beginning years, "people didn't have fences, just open spaces between houses and dirt streets." Since then in the 1990s the City of Los Angeles became the owner of Bolton Hall and the adjacent park and retrofitted the river rock building according to earthquake building codes. Prior to this the structure experienced several sizable earthquakes without damage, not even cracks in the walls "from the '71 earthquake," Judy said. Since the city retrofitting she mentioned that "our building has several cracks since the repairs and there were none before." Possibly and probably the stone building was made too rigid and being restrained to movement cracks appeared in the normal building's expansion and contraction from daily temperature changes.

"Why do you think stone was used instead of wood to build Bolton Hall?" I asked.

"Because it was here, that is what was available," she replied.

Consider also that George Harris was the builder,

a mason not a carpenter. The whole area is decomposed granite, full of granite stones and being near a river, the stones were rounded. George Harris studied the rocks and he would "fit them tightly...in such a way that you would not see the mortar." Every stone in the wall every mortar joint every opening he thought about as an artist considers his painting. Some may think it is easy to stack stones in grout. But in knowing the art's refinements lies the difference between failures and lasting buildings.

"The big draw to Tujunga was the air, the clean air. You moved here with your asthmatic kid and with this clean air your kid recovered," Judy mentioned. At the turn of the century, tuberculosis affected many people in the eastern states, "It was in *Coronet* magazine," she continued, " that this was the best place to go for asthmatics and T.B. people. The air was especially clean. Air out of the canyon keeps the air clear." Tujunga, a Shoshone name meaning 'place of the old woman' refers to Mother Earth.

During our visit in 2018, this multi-generation neighborhood remained quiet on this sunny weekday morning; it remained true to its past—unhurried and peaceful.

Sunland, Tujunga, La Cresenta, and Montrose, these

3722 4th Street, La Cresenta, a simple home of around 600 square feet remains well cared for today.

In 2017 while photographing the La Cresenta homes we saw this building site and noticed the abundant stones excavated at the new foundations.

Left and above: ***The cobblestone home remains firm and stout while the thin cobblestone wall of more recent vintage occasionally drops a stone onto the neighbor's drive.*** Below: ***3747 4th Street, La Cresenta and neighbor's stone home.***

2455 Orange Avenue, Montrose. Two stone homes on this land built in the early 1920s. Above: *And a few had stone walled basements, definitely not a SoCal feature but familiar to the mid-easterners. This one was dug out and built a few years after the stone home was constructed in the early 1920s.*

Above; ***This beautiful home at 5217 La Cresenta, Montrose received conscientious care over its century of life showing its owner's pride.***

Right; ***And a more basic rock home that received less sensitive changes over the years.***

Above right; ***This old photo of a cobblestone store came from Patricia Green's Facebook page.***

6928 Shady Grove Street, Tujunga two bedroom one bath, 1,100 sq.ft. And a more basic rock home that received less sensitive changes over the years.

Tujunga United Methodist Church built in 1919. It was the first church in Tujunga.

Early Tujunga photo shows the alluvial fan the town was built on below Tujunga Canyon. Seldom-seen snow covered the San Gabriel Mountains and shows us that a much wetter weather existed then.

communities are grouped together like pieces of a jigsaw puzzle but have a couple things in common. They are all in Los Angeles City and each has its unique group of cobblestone homes. Given that the boundary lines of each interlock and meander over the alluvial plane landscape of La Cresenta Valley, at times even new residents may not know which community they live in, as such the authors may have incorrectly listed a home or two.

This valley holds the largest collection of Rockitecture in California. Why I wondered? The land definitely has millions of stones as does other foothill communities such as Duarte, Glendora, Claremont, and these towns have a magnificent collection of stone homes but not nearly the amount as were built in Cresenta Valley. It could be that the settlers of this valley were employed in the fast-growing neighboring cities of Los Angeles and Pasadena, while the other communities lied too distant for work. Other reasons definitely exist but we found no explanations. We'll lay that on the doorsteps of historians to research.

La Cresenta Valley holds enough Rockitecture to require two to three days of touring to see them all. And many are worth visiting while a few are shielded by vegetation.

4334 BRIGGS AVENUE, MONTROSE

George, Robin, and their four children live in a simple stone home. which one source dates at 1932 and another source says it was constructed in 1903 as a hunting cabin (Yes deer, rabbits, mountain lions, and an occasional bear roamed unmolested on this foothill land.)

The street reminds us of its long history for it is named in honor of Dr. Briggs, the founder of this area. He came from the eastern U. S. to help his patients of tuberculosis and lung ailments, and he founded and developed this area of Cresenta Valley in the late 1800s for its clean, dry air in assisting his patients.

When we visited six cars were parked on the side of the large front yard, three of them classics. As Scott and I

2729 Community Street, La Cresenta.
A young family enjoys their home in the higher foothills. Chris and Ashley maintain it with pride, though not aware of its historical significance. Built in 1925, details such as the cement caps on the rock walls at the stairs and front porch, as well as the stone's tight joints hint at the wealth of the original owner. Bottom left: ***The well was 200' deep and is capped off, built by the original owner of the home to source water before utilities were installed. One can imagine that this home was among the first built in the area.*** Botom right: ***These are two of the giant boulders in their front yard. It's doubtful that any of the prior owners found these in the canyons and brought them here; it is more probable that mud flows from torrential rain washed them down and they were left here, too big to move and not worth the effort.***

Cresenta Valley, 1914, looking westward. Note the dark clump of native trees on the left side of the photo at the base of the Verdugo Hills. These appear to be California oaks growing in the low level land. Telephone poles extend along the side of the dirt road, which probably became the current Foothill Boulevard extending in east/west through the valley connecting it to Pasadena on the east and San Fernando Valley on the west. Source: USC digital library

photographed and admired the home, three dogs trotted over to us from the front yard with greetings of smiles and wagging tails.

George stopped working on one of his children's car and walked over to us happy to say hi as if we were old friends from high school. After introducing ourselves, we talked about our interest in river rock houses. But being a car guy, I commented on the cars, the classic and the four wheelers ready to play in snow or sand and George mentioned that "our kids love those cars."

As we car talked I noticed that the home looked young. It received thoughtful rejuvenation over its decades of life. A large single-story addition attached to the rear of the stone home, most likely for the kids, and the recent front French doors, trim, and porch posts brought new life to the century-old home. George is a Chiropractor, and his years of practice show in a quiet physical and patient strength. His wife Robin invited us in to join her in their living room while she found scrapbooks on the historical lineage. With pride and a special love, she told us of the newspaper articles and pictures on the home's Cresenta Valley Historical designation.

The article that Robin showed came from the *Glendale News Press* describing the area's history and the special nature of the home.

This idyllic setting is at 7388 Valhoo Lane, Sunland.

7388 VALAHO LANE, SUNLAND

In the 1970s Richard was house hunting. A do-it-yourself man in his late 20s, derelict buildings did not disturb him. His realtor warned him not to be scared off by the home he was going to show Richard. Not a home, more akin to an abandoned WWI vine-covered bunker in the canyons of the Verdugo Hills. Those small portions of walls not overgrown with unkempt vines showed this building to be two stories of rounded rock walls.

He took the challenge, and four decades later he had completely restored the old summer home. We met Richard gardening and fixing the irrigation on a sunny spring afternoon. Located at the south side of Cresenta Valley in the lowest elevation of the valley, water flowed in streams and California oaks thrived. Now centuries old, their shade continues to cool the canyon and shelter the home from the summer sun.

Richard researched his find and, after clearing off the vines, discovered initials scratched in mortar: "G. H." George Harris of Bolton Hall fame built the two-story home in 1921 as a summer home for a wealthy family

who lived in Cresenta Valley. Its two bedrooms and two bathrooms and two stories are indications of wealth along with being a cooler second home for the hot months.

Earlier we discussed the performance of these building during seismic events. Richard told us of his experience during the 1994 Northridge earthquake. While gardening he felt the land rumble and roll, he looked back at his home in fear of damage. Instead he witnessed the unexpected reaction of the stone home to the quake. Like a sail boat in ocean waves, his home rode the earth waves without damage, no cracks, no loose or fallen stones, no broken glass. George Harris knew how to build these stone structures. Monolithic like a three-dimensional jigsaw puzzle, these stone structures ride ground waves with ease.

2545 HONOLULU AVENUE, MONTROSE

In 1984-1985, this home became converted to the Montrose Montessori School. The side of the stone building shows that the city required seismic changes as seen in the wide smooth grey concrete under the arches. However, before these arch supports were installed the building went through a number of earthquakes without collapsing or cracking while this was a home. Drive east on Honolulu to visit the tree-lined Montrose neighborhood stores, an inviting quiet commercial strip of low profile stores and restful pedestrian atmosphere. An oasis of civility in the chaotic land of Los Angeles.

The original Blarney Castle in Ireland

BLARNEY CASTLE, 10217 N. TUJUNGA CANYON BOULEVARD

The inspiration for this stone masterpiece resides, of course, in Ireland. A businessman of Los Angeles had this, his dream castle, built in 1919 of 1,720 square feet with two bedrooms and two bathrooms. Two years after he built his vision, Dr. Virginia Tenney Smith bought it for her home. Dr. Smith was the first female doctor in Tujunga area and she created a botanical masterpiece

Blarney Castle

on her land possibly to grow and research healing herbs. In January, 2006 it became Los Angeles Cultural Monument #830.

An interior room of an office and library. This room is two stories or so high as evidenced by the door and chair heights.[4]

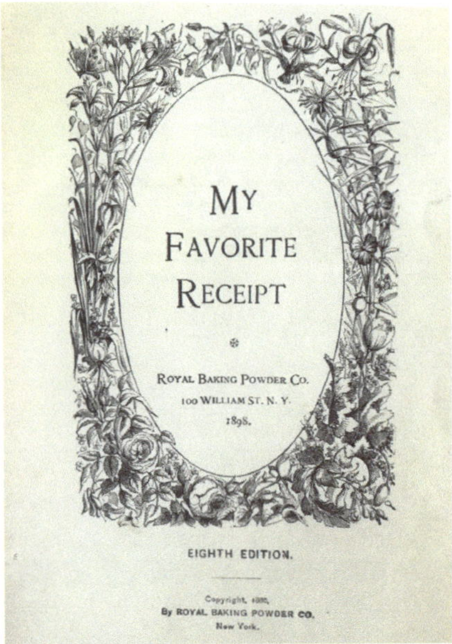

An 1898 book of recipes, but recipes were named 'receipts' then.

8436 HILLROSE AVENUE, SUNLAND. THE ROWLEY HOME.

"It was 1882," Cecile Vargo writes in the website Explore Historic California, "when the Rowley brothers wandered up to the hills of Tujunga" from Los Angeles. Riding in their wagon, loaded with bee hives, they sought land that could be homesteaded, land that the federal government granted title of 160 acres under the condition that the homesteader build a home and live on the land for five years. This land did not have electricity or telephones or water. This land was "out in the sticks" and bare. "Their first home was a tent," Vargo tells us. And that was typical for these early settlers and also for many of the latter settlers of the early 1920s.

Loron Rowley and his brothers sought a more prosperous life than was possible in their home state of Minnesota; for after the Civil War a severe Depression resulted throughout the thirty-eight states. On this homesteaded acreage the Rowleys made a living with their beehives and honey, growing peaches and olives, along with selling timber to Los Angeles residents. Yes, centuries-old timber grew plentiful in the canyons of the San Gabriels then, and the Tujunga is one of the largest canyons which extends for miles into the mountain range.

The Depression eased, and with post-war prosperity, the Los Angeles area grew exponentially and Rowley ultimately expanded the original free land to over a thousand acres in the Cresenta Valley. Also in those years he met, and after two years of courtship, married Virginia Florence in 1893.

Vargo wrote that "By 1905, the Rowley family built the first rock home in Monte Vista (Tujunga)" on Hillrose Avenue, then a dirt road named Hill Avenue. Close by and probably the reason for building at this location, "a running spring kept the ground damp and the grass green year round." Daily activities for Loron Rowley meant tending to the family farming and beehive businesses, occasionally being hired to help construct a home or commercial wooden building, while Virginia not only tended the home but also taught and took care of the "kitchen garden" where their food would be grown, and later as the children grew self-sufficient, she became the local postmistress. In the evenings after finishing their dinners, the family may have gathered on the front porch in warm summer evenings to play games, practice music, or study their school lessons. On Sundays attending church services occurred religiously, Protestant, Episcopalian or Methodist being the preferred congregations, and then socializing and enjoying a Sunday family dinner with friends and relatives.

9920 HIRONDALE, LA CRESENTA

"Rock homes are quiet," Joe Gebo said. We met him in the summer of 2018 at his beautiful solid cobblestone home named "Rock of Ages." "Cool in summer, but can get hot after three to four days of continued heat," if the night air does not cool off with ocean breezes, which occasionally occur in the summer months.

Joe and his wife Kristen's

home of 1200 square feet, with an unusually large three-car stone garage, was possibly built by George Harris in 1921, which in La Cresenta Valley gives the unique prestige to your home as if it was designed by a famous architect. Joe mentions that these "rock walls are up to three feet thick."

Rock of Ages, 9920 Hirondale, La Cresenta.

ADDITIONAL HOMES IN LA CRENSENTA VALLEY

When the lots were subdivided in the early 1900s, the intent was to size them large enough—about 75 feet wide and 300 feet deep—for the midwestern farmers to grow their own produce.

2625 4th, La Cresenta

2336 Panorama Street, south side

2301 Orange Cove, La Cresenta

END NOTES

1 Larson, D., Docent Handbook, Little Landers Society at Bolton Hall, 1983.

2 Knight, A., "Stonehurst Survey," Draft, April 2007, p. 33.

3 Harris, G., Letter, archived at Little Landers Historical Society, Bolton Hall, undated.

4 Photograph #24593, Security Pacific National Bank Photo Collection, 1921, tessa.lapl.org.

CHAPTER 17

Return to St. Lukes of the Mountains 2018

"We shall not cease from exploration
And the end of all our exploring
Will be to arrive where we started
And know the place for the first time."
T.S. Eliot, "Little Giddings" in *Four Quartets*

We returned to St. Luke's of the Mountains in Tujunga, and Reverend Jones continued the discussion we began in Chapter One. "Our parishioners are proud of their rock church. It is not only their spiritual home but also the center of their community. On my end of things, funds and people are always available to maintain our church. I think that's the way it should be with all historic buildings."

We've traveled far, never leaving Southern California, to discover these river rock buildings, and their stories of people seeking self-sufficient serenity and new beginnings. They took advantage of the eons of volcanic and seismic activity and the weathering of granitic mountains. Our search revealed the timeless depth of life, a peek into hard labor, dreams, and beauty in the hands of artists.

Seeing the original oil painting in the rector's office, I understood the parishioners' attraction to the building design and its textured solid magnificence. This church became the product of people who cared about the quality of their work, expressed community pride, people who sculpted their religious beliefs into a stone monument, one loved by its congregation today.

Churches were a fundamental part of these settlers' lives. They built rural temples and churches, local Protestant, Catholic, and Jewish.

Seymour Thomas liked to paint small landscapes outside his studio at "Cuddle Doone" or in the surrounding countryside as a means of relaxation between portraits. He often gave these landscape paintings away as gifts, some signed informally, some not at all, which causes confusion in today's art market as Thomas's work becomes more valuable. (Courtesy Witte Museum.)

The houses of worship also provided space for entertainment, such as dances, amateur plays, hypnotism acts, Christmas parties, and an occasional concert. Live performances and activities.

Churches do "much to shape the moral tone of the town," Edith B. Bowen wrote, "when the people looked almost entirely to the churches, not only for religious services, but for public entertainment as well."[1]

Many stone structures were built much earlier than

The only plans for building St. Luke's of The Mountains was this painting of a village chapel in England by Seymore Thomas.

our Southern California buildings and the most renowned and grandest resides on the edge of Cairo. The oldest, most enigmatic stone structure stands formidably outside of Wiltshire, England. I'm reminded of these ancient monoliths for their common aspect with our California stone homes and churches. All were not built for money (except for the Stonehurst tract in San Fernando Valley). One was built to worship God, the other built for a deceased god, and these SoCal river rock homes were built for immigrant families, with their churches as anchors for the new communities. All stand to this day

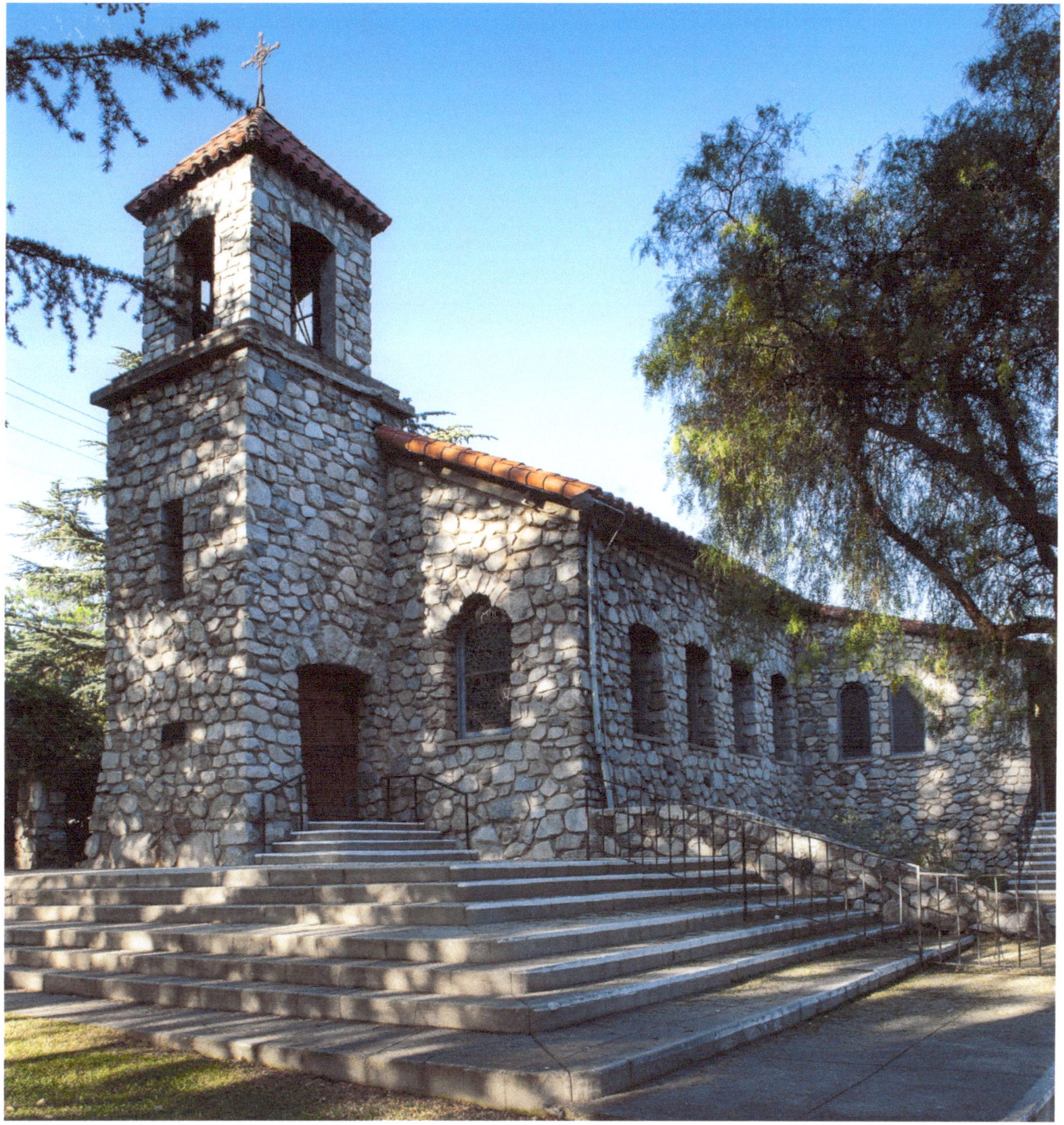

Various views of St. Luke's

and will remain for centuries more, unless demolished by ignorance or carelessness.

These buildings of *Rockitecture* remind us of nature's complex simplicity and how far we've wandered from its solace and vitality. Living or praying in a work of art energies and inspires us while being humbled by the artist's inspired yet humble creations. Having concluded our search, I reflect on the undulating salt and pepper waves in granite and see the stones' splendor wrought by time beyond understanding. Cobblestone buildings present a spiritual quality to me, and maybe also to the masons guiding them in their artistic placement of each

A couple close-up scenes from St. Luke's of the Mountains showing the aesthetic of cobblestones.

St. Luke's of the Mountains, Tujunga.

Pitzer's home at 4353 Towne Avenue, Claremont.

stone. This is so different from the dull monotony seen in bland stucco buildings. The unique quality of nature's cobblestones refreshes us from our daily routines.

Many uninspiring buildings had the decency to collapse from various earthquakes; yet, the designers, owners, and cities haven't learned from the failed buildings and rebuilt with equal structures, and somehow exceeded the visual blight, possibly from indifference by the local bureaucrats and undemanding or dulled residents. This blandness and chaos highlights and makes uniquely special the river rock homes and churches. Rockitecture today stands out when you slow down.

Though many cobblestone structures succumbed to untrammeled development as Albert Knight notes in his study, "Unfortunately, far too many examples in the western part of this region have been allowed to decay, or even be destroyed."[2] Knight refers to the stone homes that were in Los Angeles County and Los Angeles City. It is at these locations that we began our journey years ago. Despair at a seemingly bleak situation enclosing our river rock homes blossomed into joy and confidence that they will continue to be preserved with pride. We found brilliant and enlightened practices of respect in cities such as Claremont, Upland, Riverside, Pasadena, and South Pasadena. Immune to the rush of the urban city pace, these renewed cobblestone stone buildings bring the past into the present and continue to lead communities in recognizing and admiring, respecting, and preserving their beautiful cobblestone history.

"Hopefully this can invite people to stop and think about what these homes mean," Beatriz Cortez said in an interview referring to the stone homes. Her comment is salient to our presentation. "People just walk by the houses, and take them for granted," and it would be educational and fun "thing for people to think about the materials" they used to construct their river rock homes. Respect. For during the years we devoted to uncovering the context and the stories of our stone structures, we found consistently a theme of pride, and yes respect by the communities, the owners, and the parishioners.

We began this journey to locate the cobblestone buildings of California and their discovery has transformed us not only with the knowledge of geology and construction techniques, but also their sadly brief period of construction. I became inspired and thrilled by pride of the owners of these rare pieces of Southern California history. I encourage all who can to tour these communities to see these stone buildings, as their grandeur comes alive in person.

The appendix contains a listing and locations of many SoCal cobblestone buildings. Respect the vacant ones, admire the restored and occupied ones. Print a map of your area of interest and enjoy visiting these cobblestone works of art and their secluded communities, which may invite you into the less frantic world of yesterday.

Rockitecture II will extend our search into the Mojave Desert, the Antelope and Owens Valleys and further north for there we find more of California's abandoned and renewed cobblestone buildings, some with unique stories and history.

END NOTES

1 Bowen, E.B., *Annals of Early Sierra Madre*, Sierra Madre Historical Society, 1950, p. 125.

2 Sesma, C., "Historic Stonehurst District," *The Valley Journal of San Fernando Valley Historical Society*, vol. 67:9, September 2017.

GLOSSARY

Rockitecture Registry

The author's original worksheets of the river rock buildings.

SITE and LOCATION	# OF BLDNGS	HISTORIC SIGNIF	PRVT	PBLC	NOTES	YEAR
ALTADENA/PASADENA/ARROYO SECO/ SIERRA MADRE						
Farnsworth Park, on N. Lake at top, 568 E. Mount Curve Ave. E, Altadena, CA 91001	1	X		X	community center north and west sides along with fireplace and small park ranger house. On line lookup: William D. Davies Memorial Bldg., Altadena, good history there	1934 under WPA
2500 Altadena Dr., Altadena 91001	1		X		beautiful home, facing west and south, 2,000 sft., 2/2. Ltr sent 10/12/16. ' Wilkins house'	1929
Bolt house 395 S. Grand, Pasadena (Cobbleoak)	1	X	X		first floor of 2 story mansion. looks like veneer stones.	
Oak Lawn Ave, sidewalk portal, South Pasadena landmark #9 historical landmark	1	X		X	Greene and Greene style.	
Oak Lawn Ave. Waiting Station @ SEC Raymond Hill and Fair Oaks, South Pasadena	1	X		X	designed by Greene and Greene, station originally built for passengers waiting for the big red cars on the pacific electric rail line.drwgs are on line. Another source noted that T.W.Parkes designed the station in 1902	1908
314 Del Rey Ave., Pasadena 91107	1		X		faces west and south, seen on corner of Sierra Madre Blvd., garage too, good condition.	1929
485 Madeline Dr., Pasadena	1		X		can't see for foliage	1945
2912 Paloma, Pasadena, just east of Sierra Madre Blvd.	1	X	X		3/2 1900 sf.,	1923
1751 Rose Villa, Pasadena 91106	1		X		beautifully maintained, faces south and west on the crnr of meredith st, a true rock house, rsch history; note sent 6/5/17	1924
Busch Home, Pasadena	1	X			demolished but can see photos of this magnificent estate in Pasadena Historical Society and on line	1910
Santa Anita rentals just below Santa Anita dam, above Sierra Madre	81	X	X		drive up double dr. to Chantry flats parking, walk towards Sturdevant Falls, down into Santa Anita Cyn, Talked with owner of #58, Treavor Eyster, cabin #17 built by master stone mason Al Foulds, compare quality w/ DIY cabins like #50	1905+/_

so pas watering trough, Located in the parkway on Meridian Avenue between Mission and El Centro streets	1	X		X	South Pasadena historical landmark	1880's ?
BURBANK/GLENDALE/SUN VALLEY/SHADOW HILLS						
1042 E. Walnut, Burbank	1		X		front section of stone, added a second floor.	1937
711 Country Club Dr., Burbank	1	X	X		in cyn, and remodeled with 2nd story added. Per listing originally built by eddie leveque, akeystone cop actor. Up Olive to end which is Country Club Dr., see my photos.	1924
across st. from 811 Country Club Dr., small rock garage + studio, walk up dirt drive to see	1				subdivided 1921, see historical plaque	1921
906 Country Club Dr., Burbank	1		X		beautiful home 2/2, 2000sft.	1928
902 Olive St. @ 9th, Burbank	1	X	X		beautiful home 3/1. 1480 sft.,	1921
906 East Olive St. @ 9th, Burbank	1	X	X		4/2 1800 sft., added onto	1922
1042 E. Walnut, Burbank	1		X		good condition, recent addtn in rear, faces north, hse could be stone on wood frame., good stone work	
2545 Honolulu St., Glendale, Montessori school	1		X	X	good, 1923, faces NE, Montessori- good seismic retro at arches in front and side, detail of fanned stone vertically placed above each arched opening- this is original. Front porch was added shortly after original building.	1923
3070 Honolulu Ave.,Glendale	1		X		well kept, faces north, 1,829 sft. 3/, major additions to original rock hse	1922
3620 Angeles St., Glendale	1		X		good, 1931	1931
Le Mesnager Barn, Winery, 3429 Markridge road, Glendale in the Deukmajian Wilderness Park, Glendale	1		X		called the Stone Barn Vineyard	1911
3751 3rd St., Glendale	1			X	south facing, 3/2 2,100	1930
3847 4th Ave., Glendale	1		X		2nd story is wood framed dutch gamble	?
3843 4th Ave., Glendale	1		X			?
3822 4th Ave., Glendale	1		X		good home, faces south, renovated in 1969	?
3817 4th Ave., Glendale	1		X			?

1515 W. Kenneth Rd., Glendale	1		X		very good, faces south, 2/1 1,479 sft.	1926
10347 Whitegate Ave, Sunland 91040	1		X		1890-1920 blt, blocked by fence in front, beautiful sml hse,photo from the side, per zillow living room is exposed stone and fireplace is excellent	1890-1920
10603 Penrose Street, Sun Valley 91352 (north of Burbank)	1		X		faces south, built 1917, see file for more info, Model A's in yard good period photos, sent note 5/6/17, owner In 2000 (Mr. Christopher) very knowledgeable about stone houses.	1915-1917
13024 W. Wentworth, Arleta	1		X		some shrubs, adjacent to conrete wash leading up to Pacoima resevoir and Verdugo Hills, 2 story, rock over wood	?
8919 Helen Ave., Sun Valley					small structure seen from alley	1919
10004 Clybourn Ave., Shadow Hills	1		X		beautiful single story home on knoll, west and south sides seen from street, built 1913, one of the earliest stone bldgs. built in the area. Knight's source mentioned it was built in 1928	1913
9955 and 9983 Johanna Ave. and 9968 Sunland, Shadow Hills, Ca 91040, Vienna Restaurant	1		X	X	Augurst Furst Castle, two rock houses gone, Vienna Gardens restaurant is open, originally a public club w/ pool, parties, zoo with tigers and monkeys	1936
AZUSA/CLAREMONT/ UPLAND						
1003 Soldano Ave., Azusa	1		X		very nice home, 3/3 wit addition	1924
426 N. Dalton, Azusa	1 long one				a long wall associated with the catholic church,	?
1992 Indian Hill Ave., Claremont	1				vacant, derelict for sale located by freeway	?
369 S. Mills	1	X	X		was built as garage/workshop, now converted in 50s to home	1938
343 S. Mills, Claremont	1	X	X		Part of Russian Village, see wkipedia entry on the Russian Village	1933
3530 N. Padua					shielded by foliage but a rock home	
Pitzer House - now Sunrise Assisted Living Center, 2053 N. Towne Ave. ,Claremont	1	X	X		Sunrise phone - 909 398 4688, 3200 sft. 1910-12, Robert Orrarch architect, rehab 1984	1910-12
4404 N. Towne, NEC of Baseline, Claremont	2		X		old rock resevoir for orange groves is next door and next door to that is small rock house on east side, part of Pritzer groves	?

926 E. Baseline Rd, Claremont	1	X	X		Pitzer Ranch foreman's home. Library of Congress (loc. gov under Lee Pitzer)	
16th (Baseline) at Benson SEC, Montclair	1		X		restored pump or electric bldg., faces N, E, and W, part of Pritzer Groves,	?
Russian Village, Claremont, 306 & 339 South Mills	5	X	X		about 5 rock hses in this historical district.	1930s
N. Mountain at Mountain Lane, upland San Antonio Park at top of Mountain is park with a miniature stone castle fenced in	1	X		X	north side of the San Antonio Park is the waiting station built by William G. Kerckhof (1856-1929) for the San Antonis and Ontario Electric RR that ran in the middle of Euclid.	?
2393 Mountain Ave, Upland	2	X		X	home w/ rock garages, originally a citrus estate manse, burned in 1988, then rebuilt by current owner.	1906
Baseline just west of Padua on south side	1		X		great condition, old pump house for orange groves, faces all around	?
Baseline just one block east of Mountain, south side	2		X		good home and pump house? Facing NE and W	?
808 Baseline, Claremont	1		X		great condition, 1 story rock home, can be seen also from south side along Huron St., SWC of Baseline and Mountain	1922
618 Baseline	1	X	X		just west of Indian Hill, southside, good 2 story, faces north, was housing for citrus grove workers, good	1915
1333 N. Mountain	1		X		2 story perfect condition w/ 2 car garage, faces east and north	1907, reno-vated 1912
1108 E. Miramar	1		X		a family home for decades	?
3232 Padua Ave. and E. Miramar Ave. SEC	3		X		drive east down the private street for two more hous-es. 3,573 sft, 4/4,5	1929
Pump Houses on 873 and 932 Palmella, next to 918 Palmello, Claremont	2	X	X	X	good and in use for wells	?
Claremont Club, 1777 Monte Vista Ave.	1		X	X	909 625 6791, exercise club next to city hall, burnt sienna rock with additions.	?
Baseline (16th St.) and Monte Vista (Padua), Claremont, SEC in housing development	1	X	X		In center of the residential development is the rock bldg. restored to club house	
2314 Glenway	1	X	X		stone home with additions in other materials, blocked by foliage in front. Built by Chas. Glenny, stonemason	1929

1991 Indian Hill, Claremont	1	X		?	derelict stone home next to 210 Frwy	
SWC Indian Hill and Foothill (Route 66) , 404 W. Foothill Claremont	1	X	X		flower shop in great condition, renewed by current owner who has historic photo collection.	
UPLAND						
Stone Castle Powerhouse, 2453 North Mountain	1	X	X		abandoned, derelict	1905
581 N. Euclid, Upland	5	X	X		5 plex bungalow court style., good condition	1927
25 W. 17th St., Upland	1		X		2 story in perfect condition plus 2 story garage up alley. This location on Euclid St. 4,242 sft., 3/3	1905
4436 N. Towne, Upland	1		X		directly on narrow sidest., some vegetation	?
CALABASAS AREA						
Rock Store, 30354 Mulholland Hwy., Agoura Hills	1	X	X	X	Restaurant on the biker route through Malibu Canyon	
Strauss Ranch at Mulholland and Troutdale Rds. 30000 Mulholland Highway Agoura Hills, CA, 91301	1		X		refer to ineterview notes, the folder is in Los Angeles file as only building in Callabassas. Main home burned in Malibu Canyon fire. Walls survived but all wood parts gone.	mid 1920s
IRWINDALE						
Our Lady of Guadalupe Mission, corner of Morada and E. Arrow Highway	2	X	X	X		
La Creasenta/Tujunga/Sunland/Montrose						
visit the Verdugo Hills Cemetary (originally ‘Peace of Hills’) for history					1/2 block below Hillrose off of Pinyon ave.	
3751 3rd St., La Cresenta	1		X		south facing, also same address on 4th Ave., beautiful stone work. 3/2 2,136 sft	1930
3707 4th St., La Cresenta	1		X		stone garage, good. 3/1, 1,700 sft.	1927
3722 4th St., La Cresenta	1		X		sml hse, flush mortar w/stone face, stone on wood frame	1926
3735 4th St., La Cresenta	1		X		good home faces south, garage is rock, 2nd floor seems added.	1927

3747 4th St., La Cresenta	1		X		good, 1926, faces SE, full stone walls, 2/2, 1,100sft.	1927
3751 4th St., La Cresenta	1		X		good stone work, south facing, from 2/7/16 tour w/ Stuart Byles, blt 1927	1927
3829 4th St., La Cresenta	1		X		high roof for 2nd floor of wood frame with rock below	
4146 La Cresenta Ave., La Cresenta	1		X			
9810 Wentworth, Lake View Terrace	1		X		great smoke house in rear behind the newer frame house.	
11025 Foothill, Lake View Terrace					behind rec bldg at the equestrian center., check on history	
9920 Hirondale Ln., La Cresenta	1	X	X		small home maintained very well w/o additions, original. Known as "Rock of Ages", of coninued heat, 1200 sft. With 3 car river rock garage, possibly built by G. Harris, in register of historic places, adjacent to a concrete channeled seasonal river. Home's rock walls are 3' thick.	1921
2511 Los Amigos, La Cresenta	1		X		great condition, 4/2, 2332 sft.	1924
2613 Montrose Ave., LaCresenta	1		X		not well seen, foliage hides the home	
4317 Briggs,La Cresenta/Montrose	1	X	X		built in 1903, small house w/ small stones.	1903
4334 Briggs Ave., La Cresenta	2		X			1903
(Hidden Springs), Paradise Ranch in Big Tujunga Cyn, de Milles former hideout; at Ahmanson Road at 8955 Gold Creek, Sylmar, drive through rock gates to back of the campus and on a narrow dirt road to end.	1	X	X			
7024 Beckett St., Tujunga	1		X		stone seems spaced wider, showing inexperienced mason? beautiful maintenance, from Survey L A publication:Excellent example of residential arroyo stone construction in Tujunga. Arroyo stone houses are a particular building type in this area of the San Fernando Valley; in the teens and 1920s many residences were built here using local stone.	1925
9936 Redmont, Tujunga,	1+		X		and more stone work up the cyn., 2/1, 867 sft	1921

2729 Community Ave., La Cresenta	1		X		Tour the neighborhood, the original owner seems fairly prosperous from the details of const.: cement wall caps and detail, has a stone well on site some 200' deep, capped off. Home is 2/1 and 1,176 sft.	1925
4416 La Crescenta Ave.	1		X		Michael & Maria shared the history they found. 3/2, 1854 sft.	1914
2123 Crescent Ave., La Cresenta/Montrose	1		X		Duplex #2 blt 1945, original is 1,139 sft., 3 rms	1921
7570 McGroaty Terrace, Tujunga McGroaty Arts Center	1	X		X	Built by Arthur B. Benton, also called Rancho Chupa Rosa, Tujunga, lacm #63, see their web site "our history"	1923
St. Luke's of the Mountains, 2563 Foothill Blvd., La Cresenta, 818 248 3639, M-F 9:12, Sun 9-2	2	X	X	X	at Rosemont St., Sun Valley youth arts center, on SE side facing SW, listed in Historical Society of Cresenta Valley	1924
10620 Samoa Ave., Sunland, CA	1		X		1923, Elmer Revis home, faces west, beautiful, 2 story vine covered. See file photo	1923
7388 Valaho Lane, Sunland, CA	1	X	X		beautiful two story, oak tree shaded facing north, garage was probably the first floor, base of canyon, 2/2, 1,232 sft George Harris built it, Owner found GH initials in mortar.	1921
Bolton Hall, 10110 Commerce, Tujunga	1			X	Excellent museum and rock building	
8642 Sunland Blvd., Sun Valley, 2/1	1	X			very good, faces sw, LACM #644, revisit when open for interior and good ext. ck for times when open, see LA Historical Monument site for info, in a "high wind velocity are" per LA city	1925-6
10653 Russett Ave, Sunland-Tujunga, CA	1		X		wonderful small home	1926
19647 Russett, Sunland	1		X		wonderful small cottage,	
2516 Kemper Ave., La Cresenta	1		X		google earth doesn't show this st. street is one block north of Los Amigos St. and few houses east of Rosemont Ave, foliage and fence in the way	
2508 Mayfield, Ave., La Cresenta	1		X		faces north, renewed home, good	
7808 Kyle St., Sunland	1		X		fine little house, faces north, 1/1 688 sft.	1924
2511 Los Amigos, La Cresenta	1	X			4/2, 2,332 sft., 9 rms	1924
2427 Los Amigos, La Cresenta	1	X			3/1, 2,134 sft.	1923

9920 Hirondelle Lane	1		X		very nice larger home. 2,073 sft., 3/2	1921
2623 Montrose Ave., La Cresenta	1		X		five houses in a row on north side of busy street, 1,056 sft. 3/1, addresses 2623, 2625, 2629, 2637, 2643, 2645	1925
2629 Montrose Ave., La Cresenta	1		X		square stone tower on stone house, 1,144 sft., 2/1	1924
2637 Montrose Ave., La Cresenta	1		X		old wash is some 50 yards to east, source of the stones. 830 sft., 2/1	1924
2643 Montrose Ave., La Cresenta	1		X		1,633 sft., 4/2	1925
2653 Montrose Ave., La Cresenta	1		X		1,633, 4/2	1925
8436 Hillrose St., Sunland	1		X		well maintained, beautiful rock home.	1905
8430 Hillrose St., Sunland	1		X		beautifully restored and maintained home with excel-lent stone work in wide gardens	1904
2122 Waltonia La Canada, at Cresent St.	1		X		on south side, beautiful house, from 2/7/16 tour w/ Stuart Byles, 3/1, 1,252	1922
2246 Waltonia, County section of Montrose	1		X		2,922 sft., 4/2	1922
5026 Castle Rd., La Canada	1		X		3/3, 2,624 sft.	1932
2336 panorama st at Briggs Ave, La Cresenta/Montrose	1		X			
2301 Orange Cove, La Cresenta	1		X		excellent rock castle, 5/3, 2,467 sft.	1925
2455 Orange Ave., Montrose,	1		X		two homes on this parcel and heavy foliage	early 1920s
2458 Orange Ave., Montrose	1		X		the second home	early 1920s
2650 Piedmont, Montrose	1		X		covered w/vines, faces north; 2/1 956 sft	1924
2643 Manhattan, Montrose	1		X		3/2, 1,200sft.	1923
5127 La Cresenta, at Frances St., La Cresenta, in back	1		X		set back from st., well maintained, long drive to house, addition on right side. 4/1.75, 2,057 sft.	1918
2926 Henrietta, La Cresenta - Montrose	1		X		beautiful rock home original one story with wood second floor addition. Good condition	1926
2943 Henrietta, LA County	1		X			
3045 Gertrude Montrose	1		X		south facing, 4/3, 2,792 sft.	1936
2471 Altura Ave., Montrose	1		X			1903

9934 Tujunga Canyon Blvd., Tujunga	1		X		see notes from source, faces west, flared base and good sized. 1,718 sft., 3/1, met for a few hours with Richard see transcription	1919
10314 Pinyon Ave., Tujunga	1		X		foliage in way, doesn't look like A. Knight's photo.	1923
10423 Fairgrove Ave, Sunland, CA	1		X		nice house, wide open, Cora Corrigan cottage of 548 sft	1908
10212 Fairgrove, La Cresenta	1		X		grounds need a lot of work, double wide lot.	1930
10224 Fairgrove, Tujunga	1		X		home's original rock garage is next door to the south. Been subdivided.	1
10334 Mountair, Tujunga	1		X		the original been remodeled and added to with stone veneer to match the original. Beautifully done. The address might have been 10342.	1915ish
10402 Mountair Ave., Tujunga			X		highly shielded by fence as is the one across the street and up a rise with the terraced rock walls and weeds. Very difficult to see and good egs of those that are not well maintained. Next door is a vacant lot with foundations dug for new home. Showing the condition of the soil. Soil is alluvial fan with some stones not fully rounded as would be found in a river bed but granite and variously shaped rocks of varying size, with a smaller non useable size for building, showing that rock houses in this area required much digging and searching-hauling to get the adequate number of bldg. rocks.	1924
10354 Helendale Ave., Tujunga	1		X		front porch addition covers the home visibility, original large yet delicate stone placement, 726 sft., 2/1	1921
10329 Hillhaven, Tujunga	1		X		faces east, small home great view of valley, duplex, front stone unit is 2/1	1923
10369 Pinyon, Tujunga	1		X		3/2, 2,406 sft. A beautiful large home on the down side of the street.	1925
8651 Foothill Blvd., Sunland/Tujunga	1	X		X	old drinking fountain in park, non-functional, delicate and beautiful	
6404 Valmont, Tujunga	1	X	X		Vallentino house, beautiful and good sized, faces north with foliage blkg and dogs barking; 3/2, 2,475 sft.	1923

6928 Shadygrove St., Tujunga	1		X		built 1916, 1200 sqft, 2/1	1916
Blarney Castle, 10217 N. Tujunga Canyon Blvd., Tujunga	1		X		LACM #830, blt 1919 by a LA businessman, a stone replica of the Blarney Castle in Ireland	1919
7024 Beckett St., Tujunga	1		X		stone seems spaced wider, showing inexperienced mason? beautiful maintaince, from Survey La publication: Excellent example of residential arroyo stone construction in Tujunga. Arroyo stone houses are a particular building type in this area of the San Fernando Valley; in the teens and 1920s many residences were built here using local stone.	1925
10143 Mountair Ave., Tujunga	1		X		945 sft., 2/1. original is covered by many questionable additions	1927
7388 Valaho, Tujunga	1	X	X		perfectly restored two story cobblestone summer home, built by G. Harris	1921
LA VERNE						
2309 Third St.	1	X	X		great condition, face south and west, on corner	1927
LOS ANGELES						
El Alisal (Loomis House), Ave. 42 by Freeway, 200 e. ave 43, LA	1	X		X	lacm # 68, see info on web	
Hiner Home, 4757 N. Figueroa, Highland Park	1	X	X		LACM #105,, 1,900 sq.ft.	built 1922, 1930 per county record
4939 Sycamore Terrarce, west of Figueroa up the hill.	1		X		faces east, LACM #373 was called Fredonia St. House built for Occidental College instructors Mr. and Mrs. Gordon, the street became known as "Professors" row.	1892
811 North Bridewell St., next to Arroyo Seco Parkway	1	X	X		Beautiful cobblestone bungalow by Fray. and for sale on 7/22/20	1924

11813 Tiara St., North Hollywood	1		X		original building, but vine covered fence in front, was owned by penelope banning,' river rock cottage', may have been built for police chief of No. Hollywood area. 852 sft., 1/1 rocks dragged from Tujunga wash area.	1923
9901 Dronfield Ave.	1	X		X	in Stonehurst is the community bldg. LACM #172	1930
10145 Stonehurst	1		X			
10021 Stonehurst Ave.	1		X		pep Remp developed this and other stone homes in stonehurst. Built by Montelongo, built with arches in the home and it has the largest and nicest garage in Stonehurst.	1920s
10103 Stonehurst	1		X	X	original post office bldg. for the neighborhood. LA County Recorder has guest house being built in 1947, probably when converted	1923
11202 Sheldon, Stonehurst	1		X		original real estate agent's home	1923
11214 Sheldon, Stonehurst	10 /-		X		various stages of poorly performed remodeling for many of these little rock homes	1923
11102-11166 Allegheny, Stonehurst	20 +/-		X		Stonehurst Village homes	
1071 Wealtha St., Stonehurst	1		X		multiple uses including Stonehurst Community Service Station, grocery store, the neighborhood bar, and a private library (2001)	1924
11131 Thelma, Stonehurst	1		X		as of 2000, the owner is son of the original owner and has done a good job of restoring	1925
11306/11308 Laurel Canyon Blvd., Mission Hills Family Services, San Fernando	1	X	X		a beautiful stone house. owner the R. Keith , used stone on 2x4 studs,, currently Franciscan Sister Retirement Home 818 898 1546 per A. Knight "a Montelongo structure for sure. Magnificient. Franciscan Sisters bought it 1968. "A rare intact example of an arroyo stone house in Mission Hills..." from Historic Places LA	1930
San Encino Abbey, 6211 Arroyo Glen St., Highland Park (Garvanza)	1	X	X		LACM #106, see most compete write up at "the bohemian Browns (Clyde Browne) of the arroyo seco' KCET link, also see on youtube the promotional video, https://youtu.be/ZlyypX88hvk	1915 (1909-1925)
Lederer Res. 23134 Sherman Way, West Hills, LA	1		X		on hilltop and fenced, access restricted	1933

Lederer Stables, Canoga Mission gallery, 23130 Sherman Way, West Hills	1	X	X	X	built of sedementary ocean bed rock, LACM 135, see file, Mission Revival Style stables built in 1936 by Francis Lederer, converted into a community arts center 1970s	1936-1937
13333 W. Filmore St.	1		X		arroyo stone home in Pacoima, Tony is grandfather owner, son is Tony jr. stones from Hansen Dam and Tujunga Wash	1932
NWC Amigo and Bassett, Reseda	1		X		some foliage, set back, small river rocks, no variety in size, appears to be veneer on studs built a 1920as farm house.	32m
RANCHO CUCAMONGA						
5510 Hermosa Ave., Rancho Cucamonga	1	X	X			1930 per county record
in Alta Loma foothills					abandoned and partial standing CCC building, in foothill open land	1934
Rochester and Foothill Blvds.	1				stone home blocked by foliage.	
8308 Baker, Aldelfer home	1		X			1927
SWC Hillside and Hermosa	1		x		great condition home	
9983 Hillside Rd.	1		x		drive east on hillside might be others there.	1921
Demens-Tolstoy Estate at 9686 Hillside and Archibald	1	X	X		See wikipedia	1890
7690 Archibald Ave., north of Foothill	1	X			1907 stone church, beautifully maintained. So. Of 210 and west of 15 fys. Faces south and east.	1907
SAN FENANDO CITY and STUDIO CITY						
13170 Foothill Blvd., Sylmar	1		X		house is good, faces north	
503 Hagar St., San Fernando	1		X		highly modified, can barely see the original stone home.	
539 Alexander Street, San Fernando	1		X		parially covered with vines but a good stone home.	

539 Harps St., San Fernando City	1		X		good one, craftsman style, 1 blk s of 5th and Harding, 3-4 houses	
1219 Seventh Street, San Fernando, maps software uses '7th street'	1		X		lots of vegetation	
1223 Seventh Street, San Fernando	1		X		house covered by foliage	
4146 Klump Ave., Studio City. Ms. Pam Bam spkg of her dad Mr. Bernard Bowman	1		X		beautiful, faces west; owner had rock house built (close to la river which was not concrete channel then), wanted one similar to what he saw at Stonehurst.	1927
11308 Laurel Canyon Blvd., San Fernando	1		X		beautiful, authentic river rock house.	
RIVERSIDE/GLENDORA//ONTARIO						
Mt. Rubidoux	2	X		X	Castle & bridge on road to top of Mt. R, So. of 60, west of 91, Exit 91 @ Mission Inn Ave. see website	
2110 Mission Inn Ave., now Seventh St. and Kansas Ave., SWC Riverside see next line	1	X	X		at front small stones of even size in rows, towards side and rear larger stones were used, in both cases this appears to be stone on studs or veneer. Small one car garage attached on west side. Moderate condition now artistically built.	based on visual esti-mate early 1920s
7th and Kansas, Riverside	1		X		a gem of a small rock house with the finest of detail-ing, This is veneer, probably on stucco backing	
Indiana Ave., 1/2 block north of Jane St. on east side Riverside	1		X		poor condition, needs work, occupied	
7038 and 7050 Indiana Ave., Riverside	2		X		poor condition, front yard lost to street widening, appear to be same builder	late 1920s

Guasti Winery, Guasti, Ontario; south side of 10 Frwy, 100 yards to the north, and next to the railroad tracks	1	X	X		see web site www.boom, directions in green note book (10 e to Guasti Blvd. in Ontario, south to Guasti Rd., west to turner, left to tracks and dirt road, drive west on dirt road. Use telephoto also turn around go onto N. Turner east to Guasti Rd., then left to old stone winery on left. Area is fenced "Private Property" and inaccessible. Large stone packing house bldgs.,wood roof sags and collapsed in areas, vacant with windows boarded tight, this is the site of the town of Guasti	
on Foothill Blvd. in Glendora is an immaculate restored home on north side at a corner	1		X			
400 Glendora Ave., Glendora a dressed granite church	1	X	X	X	Church of cut granite, see B's photos	1931
Rubel Castle, 811 Live Oak, Glendora	1	X	X	X	To tour call 626 963 0419, guided tour is a couple hours, on youtube - Huell Howser	
SOURCES					**CONTACT**	
Pasadena Historical Society					651 S St John Ave, Pasadena, CA 91105, 621 441 6333	
Little Landers Historical Society					818 352 3420, little landers@verizon.net, Regina Clark-museum director, Jackie Kavas-archivisit	
Highland Park Heritage Trust						
USC Library Special Collection by Security Bank					historical SoCal photos	
USC Library Calif. Historical Society collection					213 740 5900, 9-5 m-f 2nd flr Doheny Libr	
"Dick Whittington photo collection" at USC library					use special collections, 9-5, m-f, doheny library, 2nd flr, 213 740 5900, soecol@usc.eduoral histories	
Cresenta Valley Historical Society La Crescenta, CA 91214 Phone: (818) 548-3795 Fax: (818) 248-7202						
City of Riverside Historical Dept.						
Claremont Heritage, Garner House in Memorial Park					3900 Main St., 951 826 5311, library is 951 826 5201	
Claremont Colleges Digital Library					840 No. Indian Hill 909 621 0848, M-F 9-1:30,	
Cooper Regional History Museum, 217 A St., Upland					ccdl.org	
Upland Heritage					738 W. 17th Street, 909 981 0280	

Glendale Public Library, 222 E. Harvard St., ck. Hrs. 2:30-10pm				909-981-0280, gharboe7@gmail.com
Glendale Historical Society				chck flickr site and Big Read page, "Stone cottages of Verdugo Hills" and a brochure "from stone castles to cottages", Chuck Wike, community relations mgr, 818-548-2042
Source = survey LA, LA Dept. of City Planning, Sun Valley, Tujunga Canyon Report				(818) 242-7447
Source = Images of America - Highland Park by Charles J. Fisher and the Highland Park Heritage Trust, Arcadia Publishing 2008				(843-853-2070)
Source = Tartaglia, Louis James, 1981, "Cultural Resources Survey" Tract # 41695, EIR # 53-81				arcadiapublishing.com
digital library.csun.edu/ special collections section, A. Knight file and photos				for Tierra Engineering Co., Tarzana, Ca. on file at South Central Coastal Information Center UCLA
LA Public Library, Central Branch, Calif. History collection				
LA Public Library, Central Branch, Security Pacific Bank photograph collection (in the history and geneology dept.)				Calif history is in the 970 section
DRAFT STONEHURST Historic Resources Survey 4/2007 prepared for Los Angeles Dept. of City Planning, prepared by jones & Stokes, LA				central LA library historical section
"Sunland and Tjunga " The making of America series, marlene A. Hitt of Little Landers Historical Society, Arcadia publishing, 2002				on file at Bolton Hall Museum
"Lost and Found II", Elizabeth Pomeroy, Many Moons Press, 2002				
The Historic American Bldg. Survey, fed doc. Ongoing				
Valley Sun Newspapers, Inc.				
look up "model colony" in history room Ontario City library				818-365-3111, Thelma Barrios editor in 2003
Sherman Foundation				Corona del Mar for complete historical phone books and Calif history,
calisphere.org				
Altadena Historical Society				Address: 730 E Altadena Dr, Altadena, CA 91001 Phone:(626) 797-8016 M-T-F 9-1,
Gordon Howard Museum, 115 N. Lomita St, Burbank, CA 91506				Look into Will Thrall papers and his articles on the Big Tujunga in Trails Magamine, winter 1938
San Bernadino County Museum				

Fairchild Aerial Library UCLA Geography Dept.				currator of history (maybe Michelle Nielsen)
Cooper Regional History Museum, 217 E. "A" St., Upland				for aerial photos of the areas in their respective times.
Upland Heritage				P.O. Box 1828, Upland, 91785, 909 981 0280
Historic Places LA				217 East 523 Sixth St., Suite 826, LA 213 623 2489. L A Conservancy. online resource of LA's historic resources directory by area.
Glendake News Press 4/25/2008 article by Ruth Longoria				
Center for Oral and Public History, Calif State Fullerton				scvhistory.com. 909 787 4787
Trevor Eyster				home owner in Sierra Madre Canyon, trevor.eyster@gmail.com, also look up "Santa Anita Canyon History" by Glen Owens
Stuart Byles interviews Stuart Byles: 818-249-2414, Stone Barn Conservancy				
Old Stone House website				
Ontario City Library, "Model Colony History" Room				At Ontario City Library
L.A. Conservancy				Adrian Scott Fine, Director of Advocacy

Meet the Author and Principal Photographer

Barry Schweiger is a retired Los Angeles-area architect who fell in love with river rocks during his childhood and who felt honored to renovate and refurbish Rockitecture during his career. His passion and eight years of research are infused into every page of this book as he guides us on a journey of the Rockitecture of Southern California.

Scott Clark is a lover of fishing, cars, birds, and photography. For 30 years Barry and Scott roamed California's Mojave Desert and the Sierra Nevada Mountains and saw old river rock buildings igniting the desire for this book. Then for eight years we researched, hunted, and found over 200 remaining river rock structures of California. Scott continues to travel to cold and far away streams to fly-fish.

www.ingramcontent.com/pod-product-compliance
Lightning Source LLC
LaVergne TN
LVHW070817210926
848482LV00004B/284